AirCrewStories

Real stories told by real people

Volume 1

James McBride

Kais Kassim

Published by: flightsofpassion.com

www.flightsofpassion.com

First published 2018

Second edition April 2025.

PAPERBACK ISBN: 978-0-9931368-8-7

Jacket design by Daniel E. (@twenty4hrdesign)

Preface

Just like the aviation industry, there is a global reach to the contents of this book. There are stories from as far West as Seattle and as Far East as Hong Kong. We have the inside story of an attempted hijack of a B747 in Africa. There is a rescue flight in the Balkans, an Engine Failure-on- Take-Off in France and running out of fuel in Scotland in the winter...

You know that moment in the 1980 film **"Airplane!"** where the flight attendant makes the cabin announcement, *"by the way is there anyone onboard who knows how to fly a plane?"* We have the story of one Private Pilot whose dream came true. Could he fly an airliner?

We wrote AirCrew Stories, so we could share these real-life experiences with a wider audience. Our colleagues in the industry have pulled no punches in their descriptions of the events which happened to them on their flights. The language is often colourful and occasionally basic. We make no apology for the rawness of the narrative used by our contributors. Theirs is the domain of the operational 'Front Line' and just like the soldier, facing the enemy, they tell it how it is. These are operational crew and professional aviators, often facing life or death situations, you would expect it to be gritty.

Above all however, the message which comes through is the indomitability of the human spirit. It is an industry which has demonstrated that the safest form of travel is by air and which prides itself on Safety First! In a similar way, you will be proud of how the AirCrew in these pages overcome the challenges which face them. You could say they were just doing their jobs, but by the end of the book we hope you share our view. These are exceptional individuals performing incredible feats at the top of their game. We are so proud of them all and sincerely thankful for their decision to share their experiences with us.

James McBride and Kais Kassim.

AirCrewStories

Real stories told by real people

Have you got an aviation story to tell?

Send it for our next book.

Share your career journey with the world.

Email to stories@aircrewstories.com

@aircrewstories

Facebook/aircrewstories

Foreword

I flew with James on one of my first days of Line Training for easyJet on the Boeing 737 as a new First Officer. Though it was obviously a really busy time and everything was very new, he made sure we still had a 'fun day out'. It is pleasing to see that some 18 years later, he still has the same passion for aviation. So much so, that he writes about flying in such a way that everybody can gain a greater appreciation of our profession. It is through books like this one, that the travelling public are able to see what really happens behind the locked cockpit door.

James has brought together some incredible aviation stories in this book from several of our colleagues. These will appeal to all air travellers, whether they are in the flightdeck, or the cabin. If you want to know more about flying, then I am pleased to recommend this book to you. It is I think reassuring to see that even in our technical, computerised world, when things go awry, the humans are still there to play the lead roles. It is through the actions of these well trained, committed flightcrew, that our industry remains the safest form of travel.

Tony Underwood,
Ex rugby international and current Airbus 380 Captain.

Sydney, Australia 2018

The junior half of England rugby's famous flying wingers, Tony carried on the family tradition of taking to the skies. Whereas Rory plied his trade flying for the RAF, Tony took to aviation in the commercial world, making the transition from the rugby paddock to the airport apron to carve out a second successful career.

The launch pad was a visit to the local flying school at Newcastle Airport, just a stone's throw from the Newcastle Falcons stadium. He would take up some of his former team-mates on local "cross country" and circuit flights, certainly a challenge for weight and balance when they weighed close to 120 kilos! From those formative early days he now plies his trade flying the 500+ tonne A380.

Contents

Acknowledgements

Our grateful thanks are due to the following contributors.

Captains: David Barrington, William Hagan, Dan Moore, Andreas Taxakis, Nick Francis, Steve Cross, Brien Wygle, Bob Johnson, Ali Al-Hashim, and Sandy Cameron.

Senior First Officer Jon Mytom-Hart.

Darragh Owens, Head of Training,
National Flight Centre, Dublin.

Philip Johnson PPL/IMC.

1 THE WOUNDED DRAGON

By Capt Dan Moore

Airbus A320 airliner – since 1987 over 11,500 have been delivered.

We were preparing for the final phase of our flight from Nanjing, China, to Hong Kong approximately ten minutes before top of descent, I asked Amigo (the First Officer, not his real name) if he was ready for the approach and landing briefing. He gave me a thumbs-up sign having just received the latest weather update at Kai Tak airport in Hong Kong. Typhoon signal number three was now in effect. The airport itself was affected by strong easterly winds due to the approaching typhoon, with scattered low clouds and a temperature of 29 degrees Celsius.

Runway 13 was in use. The initial approach for that runway began with a gradual descent while flying on an easterly heading known as the Instrument Guidance System (IGS). At approximately 600 feet the aircraft was required to make a right turn through 47 degrees, while continuing to descend to be aligned in a south-easterly direction before touching down. The airport itself was closely bounded by hills to the north and to the east. From a pilot's point of view, this approach provided all the excitement of a rollercoaster. It was never the same twice, particularly when it was windy.

No matter how many times I'd landed there, each time had the potential to be different. This was in part due to the terrain surrounding Kai Tak and the effect it had on the local winds. The last part of the initial easterly approach and the subsequent descending turn on to short finals took place over a heavily built-up area; the domino-like high-rise buildings planned in overlapping shapes stopped just short of the runway threshold. On this particular day, aside from the turbulence caused by the strong winds tumbling over the hills to the east of the airport, the initial approach was normal. With the landing checks completed we started the descending right turn to align the aircraft with the runway centreline.

Suddenly there was a heavy jolt and a loud bang that shook the whole airframe. It was so severe that I felt my body briefly floating upwards only to be restrained by the shoulder harness and lap straps. Despite the engine noise and sound of the wind rushing over the airframe, I heard the passengers scream in the cabin.

The aircraft had briefly encountered a combination of windshear and severe turbulence. Seconds later a warning chime sounded in the cockpit, alerting us to a failure in the wing flap system.

The windshear had also temporarily destabilised the aircraft so I elected to abandon the approach and I advanced the throttles to the go-around detent. As we started to climb away from the ground the aircraft began to oscillate violently, rolling from side to side like a feather on a stormy day, dancing randomly to the whimsical tune of the unmerciful wind gust. As the aircraft speed increased, the rocking motion started to reduce. The more the speed increased the more stable the aircraft became.

The aircraft fault detection computers warned us that the landing flaps located along the trailing edge of both main wings were jammed fully down in the landing position. This meant we would no longer be able to raise them to the up (zero) position, normally used for high-speed cruise flight. According to the manufacturer, the roll oscillations were caused by an un-optimized flap configuration.

The trailing edge flaps were jammed at full due to the sinking windshear while the leading-edge flaps were still movable by the flap-lever. With flaps full, the leading edge flaps were set at 27 degrees. As we went around, the procedure dictates to go one nudge up on the flaps, in our case we went from flaps full position to flaps three position. That meant the leading-edge devices went from 27 degrees to 22 degrees while the trailing edge flaps were jammed at full. Hence the optimization issue.

Flaps full required 27 degrees of slats and not 22 degrees and that was not optimized. At slow speeds with the heightened sensitivity of the computers due to the flight control optimization issue were compounded by the low-level turbulence caused by the strong winds of the approaching typhoon.

While Air Traffic Control assisted us with directions for our second approach to land, I discussed the problem with my co-pilot. We decided to attempt another landing on runway 13. As we began the approach everything appeared to be normal except the wing flaps were still jammed fully down in the landing position.

As the aircraft continued to slow down, the oscillating motion started again. With the violent oscillations continuing, the prospect of a descending turn at low level was too dangerous to contemplate. Simply put, the aircraft was almost out of control in the roll axis. There was no alternative, but to apply full power and go around for the second time. The erratic behaviour of the aircraft puzzled us.

All we knew was the higher the speed, the more stable the aircraft became. I also knew that any attempt to land on the south-easterly aligned runway was doomed to failure. I could not afford to have the wings rocking violently close to the ground as we attempted to accurately fly the descending right turn. Therefore, we decided to ask Air Traffic Control to use the north-westerly runway. Runway 31 had a long straight-in approach beginning out over the South China Sea to the south of Hong Kong. The only disadvantage was we would be landing downwind, which would lengthen our landing roll.

At this point we declared an emergency. We reviewed the procedures in the aircraft manuals but could find nothing to help us with our predicament. I contacted the maintenance base on the radio to seek help. I explained the situation and after a brief pause, and to my utter surprise, they came back with, "Roger, see you on ground." I knew then that even they had no technical solution for our problem. Out of desperation, I elected to engage the autopilot as I thought it might do a better job than I could of keeping the aircraft under control, with the added advantage of giving me a much-needed break! At first it held reasonably well, but as the aircraft began to slow down for landing it began to oscillate again severely.

I disengaged the autopilot and continued the approach manually. As the runway threshold approached, the oscillations became so severe we had no choice but to go-around for the third time. We now faced not only a flight control problem, but we were running dangerously low on fuel. On the third go-around, as the aircraft's nose rose above the horizon, I felt betrayed, however, I was determined to fight on, but the odds did now seem to be stacked against us.

I considered the possibility of ditching the aircraft in the sea. I asked Amigo to get the aircraft manual and to read the ditching procedure. Looking down into Victoria Harbour below I could see it was crowded with all sorts of ships, big and small. I remember saying a silent desperate prayer to myself, 'God, the sky is so violent, the earth is so stingy, and the sea is so filled with ships it is impossible even to ditch the aircraft in the sea'.

Suddenly I was filled with guilt; I had not kissed my two children goodbye that morning nor hugged my wife farewell. A plethora of self-perceived sins from my past ran through my mind, things for which I wanted to make restitution for before I went to face my maker. The co-pilot snapped me out of my self-reflection with an exclamation of his own: "We have to make it this time, WE HAVE TO!"

We felt completely alone in our plight. I have heard that people facing the possibility of death see their lives run past in a matter of seconds. For me there was a sense of an unfinished mission in this life; I needed to do more, I needed to see my young children again. The feeling was so strong it overcame all other weaknesses. There was no option, no alternative. Although I had lost confidence in the aircraft, I had not lost confidence in myself. I asked Air Traffic Control for the shortest routing possible to place us back on the final approach for Runway 31 again. I advised we were proceeding for a last attempt.

As the runway came into view I was overwhelmed by the instinct to survive, relying on my training and my past eighteen years flying experience. The aircraft speed continued to reduce. Passing eight hundred feet, the oscillations were getting noticeably worse again. Like a wounded bird limping towards its nest, we continued towards the threshold. It seemed that the oscillations were a little more controllable this time. It may have been my imagination, but the wind strength appeared to have dropped off a bit. Nearing the ground passengers started screaming again. We delayed lowering the landing gear and at about eight hundred feet, I asked Amigo to lower the gear.

The aircraft touched down firmly on the runway with a higher speed than normal. The additional airspeed had given us better control on the final approach. Seconds after touchdown the aircraft began to veer to the right and try as I might I could not stop it. The rudder pedals appeared to be jammed. We slowly departed the right-hand edge of the runway and onto the grass verge. Even the deployment of asymmetric reverse thrust failed to bring the aircraft nose back to the left.

We bounced across the grass and onto the parallel taxiway still travelling at a rate of knots. Immediately to the right of the taxiway there was a steep drop off into the stinking putrid waters of Kai Tak Nullah. Surely, we had not endured so much and triumphed over such adversity just to make an ignominious splash into the Nullah! I kept trying to steer the A320 to the left, thinking this was one heck of a way to end a career.

Suddenly, as we crossed the edge of the taxiway, the rudder pedals freed up and I could use full rudder to swing the aircraft nose left with only seconds to spare. The aircraft continued to decelerate and came to a halt on the taxiway. Although there were no signs of damage or fire; the fire trucks, ambulances and airport security cars quickly surrounded the aircraft.

Amigo and I quickly went through shutdown procedures. Noticing the rescue leader within shouting distance, I opened the cockpit sliding window. "Any sign of fire?" I called. "There is smoke coming from one of the left wheels", he shouted. I poked my head through the window and saw a plume of white smoke from the wheel area.

The rescue leader suggested we evacuate, and since he was in a better position to assess the situation, I decided to order the evacuation of the one hundred and fifty-six passengers and crew on board. The evacuation went smoothly with no injuries despite the strong winds.

As I went back through the aircraft cabin to make sure everyone was off the aircraft, I saw a crunched-up packet of Chinese cigarettes sitting on the floor. It had been three years since I'd last enjoyed a cigarette, but at that moment I couldn't help it. I picked up the pack put it in my pocket before sliding down the dark-yellow escape chute. Amigo and I were met by an anxious ground engineer who whisked us away in a small car to the airport passenger terminal.

As I walked through the terminal, I spotted some of our passengers in an incident room. They glanced through the glass windows at us with a combination of relief and gratitude. I couldn't help but meet their gaze. We had done the best we could, and they were all alive. Amigo and I went to a separate room where we were interviewed by officials from the Civil Aviation Department, followed by a quick medical check as part of the procedures for any crews involved in incidents or accidents.

The good outcome from the event was that the manufacturer changed the way it communicated important technical bulletins by color coding them according to severity, importance and impact on safety. The other good outcome was the warning that maximum autobrake is not recommended for landing and here is why. The reason the aircraft departed the runway was due to what is known as "Bearing Lock", a feature fitted mainly to assist the aircraft in a rejected takeoff by blocking the nose wheel bearing in order not to veer off to the right or left of the runway center line in case of an engine failure at high speed. That feature is linked to the maximum autobrake position only. Seconds before landing, Amigo engaged the maximum autobrake.

With the cross-tailwind, the aircraft nose was pointing to the wind away from the runway center line to the right. When the aircraft touched down, the bearing lock activated and maintained the bearing on the heading we had at touchdown. Thus, I couldn't steer the aircraft back to the runway no matter what I did. The only way to disengage it was to disengage the maximum autobrake or when the speed is below eighty knots according to the system logic.

When we did the briefing, we agreed to use medium autobrake. Frankly, even if I was aware that maximum autobrake had been engaged, it would not have changed the outcome as there were no limitations on the use of maximum autobrake.

The result was not that bad, no injuries, no damage to the aircraft, flight control modification came into effect, colour coded Operation Engineering Bulletins, computers were modified for better, and finally, new important limitations were added "MAXIMUM AUTOBRAKE IS NOT RECOMMENDED FOR LANDING."

Airbus A 320 cockpit

2 THE 'SULLY MOMENT'

By Capt James McBride

I now know how a priest feels... Maybe I should rephrase that. You see, I was in a privileged position. I was listening first-hand, one to one, as if in the confessional box to a colleague of mine talking about his 'Engine Failure On-Takeoff'. I have known Sandy for decades and we have operated airliners together as a crew in the past working for the same company. He is a true professional and I have the greatest respect for his skills. It was all the more compelling then, when we could sit together with time to talk in-depth about an event which might have shortened his life by a considerable margin...

I think it really came home to me how close they had come to catastrophe when he used the phrase, ."... there was a period of ninety seconds or so when I thought... *I DON'T think this is going to work...!"* He had preceded this by explaining that they had lost 50% of the thrust, when an engine failed, just after takeoff in 35° Celsius in a very tired old Boeing 737. Not a good situation to be in for sure. In a similar way to Capt Sully (Sullenberger) referring to the 208* seconds period after they hit a flock of Geese climbing out of La Guardia, my friend also viewed this as a pivotal moment in his career.

"Single engine, she was struggling, hardly climbing at all...that was when I thought, I DON'T think this is going to work...!"

You see 'Engine Failure On-Takeoff' (in an airliner) is THE most frequently practiced exercise in simulated training. Why? Because it is simply the most critical event which can possibly occur while flying airliners for a living. There are other scenarios which are difficult and/or hazardous of course, but to lose half your thrust, close to the ground, slow speed at heavy weight is a real tough one. Usually the aircraft is at its heaviest with a full load of fuel onboard, the crew have only just started their duty, (they are not at their most competent) and the startle factor is intense.

Not only that, but technically it is one of the hardest flying manoeuvres to execute properly. Bear in mind the physical techniques of flying airliners are usually based upon gentle pressures and smooth control inputs, but when one powerplant fails on a twinjet, rapid major response is required by the pilot to retain control of the flightpath. Use of the rudder pedals when airborne is reserved only for situations of asymmetric thrust. In the event of engine failure, (to keep the aircraft flying straight), the pilot must push nearly full rudder towards the live engine to counter the dramatic effect of the thrust from that side.

Of course as speed increases and the aerodynamic effectiveness of the rudder improves, pedal displacement can also be reduced, but in that initial stage the prompt application of rudder is vital. With the input of the rudder, there will be a natural requirement for some aileron to balance - in fact the least drag configuration will be attained with 5 degrees of bank towards the live engine.

Literally this must all happen within a matter of split seconds at the moment of engine failure, there is hardly time for conscious thought if a safe outcome is to be achieved. Imagine my old friend then, sitting there like Tom Hanks in his uniform, another day in the office... But this is no movie, this is real life. This is how Sandy pays the mortgage and feeds his kids. You see pilots are no different to anybody else; they are not superhuman, not endowed with special powers, simply ordinary men and women doing a technical job under sometimes difficult conditions.

Day to day they make decisions and calculations as part of their work which are of vital importance to the safety of their flights, passengers and crew. Perhaps in comparison with surgeons, those decisions can mean the difference between life and death, but pilots do not dwell on that part – to do so would be distracting. If you ask a professional aviator about this – "their lives in your hands?" - they may say something pragmatic like, "My job is to get us all home safely..."without highlighting the risks involved.

To do so there are countless hours of study and practice in flight simulators to ensure their skills are tested, rehearsed and demonstrated to be at such a high level they can cope with whatever emergency situation arises. Some pilots will joke about their occupation by way of coping with it, "My job is hours and hours of total boredom, punctuated by seconds of pure terror..."

That being said, most of the time of course, the aeroplanes function exactly as the manufacturer intended and the weather is kind enough to permit unaffected aviation activities at thousands of airports worldwide. In short – it's business as usual. But occasionally, very occasionally, there comes a time in a pilot's life when they will be tested to the maximum in a real-life emergency. ALL of their former experience, training and testing will be brought to bear in one short interval of time and the only thing separating the aircraft and disaster is their own personal abilities. They are suddenly thrust from being part of 'the cast of thousands' to playing the lead role in a real-life drama - it's their 'Sully moment'.

Sandy was quite calm as he talked about it to me. He was particularly praiseworthy of his colleague the First Officer. I was pleased to hear him say, "He was fantastic! Sam really did a great job. He simply said 'Engine Overheat - Number Two...!'* of course I was aware that we had an engine problem as the Master Caution system had illuminated with the amber ENG caption glowing brightly."

He took a deep breath and continued, "at first I thought he was joking, I thought; you can't say THAT to me!? You know after all those years working in the Sim as an instructor, that's what I'm supposed to say... to somebody else?" I was fascinated and listened intently not wishing to interrupt him, this was his Sully moment.

"You know the checklist memory items like I do for the 737. We don't have a choice! Well, we brought the thrust lever back on the number two engine as the drills say and the warning light went out. The number two thrust lever was now at idle power and we were flying single-engine at max-takeoff-weight on a very hot day. Sam had only just come out of the simulator, so naturally he was razor sharp with all his QRH drills and checklists. I guess he thinks it happens often that we have an engine failure on takeoff...

Well you can understand it really, it happens several times per day in the Sim, but on the line; only about once every 30 years as far as I'm concerned! He was really cool about the whole thing which was amazing, because at the time he was very inexperienced. I think he only had about 12 hours on type and this was one of his line training flights..."

I had to ask, "I guess you put out a Mayday call?" He looked at me with his head inclined to one side, his mind partially elsewhere, remembering the moment...

"Yes, we did. 'Mayday, Mayday, Mayday! <Company>737, Engine Failure On-Takeoff! Climbing straight ahead! Standby!' It was all textbook stuff really, except that the guys in the Tower hit the crash alarm and they must have had a huge siren on the wall just next to the microphone.

Every time they transmitted on the radio, there was this loud 'WEE-WAAH! WEE-WAAH!' in the background, very distracting; we could barely hear what they said! I tried to tell them... 'Can you please shut that alarm up?' But it was no good, we had to put up with it." I was curious about the effect on him, how did he FEEL?

Sandy explained, "It is incredible how the training kicks in. I know we all treat it lightly and take it for granted. I was aware my heart was pounding, but I could see we were doing a good job, working well as a team and reacting correctly. It was then as I considered the situation we were in, single-engined, climbing out at Max Weight with nearly full rudder and she was struggling... hardly climbing at all... THAT was when I thought, *'I DON'T think this is going to work...!'"*

I could imagine him using everything he knew to keep that old ship climbing away from the ground on one engine. I knew also that his passengers could not have been in safer hands, he's one of the best I've ever met. For the next minute and a half he dragged every ounce of performance out of the old girl and slowly, very slowly... he coaxed her skywards. I held my breath as he described the performance parameters which he'd had to deal with – a true 'Sully moment'.

Once he had climbed to a safe altitude - "We completed all of our checklists, spoke to the cabin and made our plan for returning to land. Of course, it was an overweight landing, but that didn't matter as it was an emergency situation. I remember thinking, 'it's a good job they haven't built a Boeing yet which can't land back on the same runway at the weight we took-off at'. That's a nice feature about the Boeing. I recall thinking; *I'm doing the famous 'One Engine Inoperative Landing' checklist FOR REAL for the first time ever."*

"So after landing and taxiing in, we got the passengers sorted out and off into the terminal. Then found hotels for everybody..." and suddenly this was Sandy the Captain talking; not Sandy - Saviour of the day! He was simply relating the other part of his job, looking after his people and dealing with the administrative functions of 'his office' – it was quite illuminating for me to observe the difference.

** Airliner engines are numbered from the left-hand side, 1 and 2 in the case of the B737 – a throwback to the 1950s. More modern twinjets refer to Left and Right engines.*

B737 No2 engine showing abnormally high vibration level.

3 THE LAST OLYMPIAN

By Capt. Andreas Taxakis

Capt Andreas after landing for the last time in Athens for Olympic Airlines.

There are many 'firsts' in a professional pilot's life. First flight at the controls, first solo, first commercial flight as a First Officer, first flight in command of a heavy transport aircraft. There are even occasions when you are chosen to carry out the inaugural flight to a new destination for your airline.

The common theme of all these flights however, is that they are joyous occasions worthy of celebration - especially the inaugural one. To be the first among all your colleagues to fly the company's colours into a new airport is a great honour. It usually signifies a new route has commenced which itself is an indicator of prosperity. The airline is expanding, the new airport will be hiring more people – everyone has a smile on their face. But 'last flights' are different, very different. To carry out the very last flight for an airline which is ceasing operations is one of the saddest tasks in the industry.

In the words of Mark S. Pyle (Captain of the final Pan American 'Clipper' flight back in 1991) "you can sense the sombre mood and pent up emotion of the crew":

"At 2 p.m. EST, the wheels came up on Clipper 436, hailing from Bridgetown, Barbados, and bound for the city of Pan Am's birth. We flew with silent thought, exchanging few words as time passed. San Juan Centre cleared our flight direct to Miami and I punched in the navigational coordinates for Miami International a final time. Little could be said in the face of a solemn reality – the certain knowledge of dead-end careers. What happened can best be described as a death in our immediate family. Pan American was my family in every sense. It was the corporate family to thousands."

An interesting fact is that this crew flying the B727 from Barbados to Miami for the very last time did not realise theirs was the last flight until after they were airborne. They knew the airline was ceasing operations very soon, but were not aware that this would be the last revenue flight.

In a similar way, Airbus A340 Captain Andreas Taxakis only learned that his would be the last revenue service for Olympic Airlines, shortly before he and his crew departed Athens, their home base. They knew then that their flight home from Toronto, via Montreal on 28th September 2009 would be the last international sector flown by the airline. In his own words, "We were only informed a few days before that ours, Olympic Airlines 424, was to be the last flight. We couldn't believe it really! We were not specially selected, just a regular roster for the flight as normal and I was not a management pilot, just a training captain on the A340 fleet." When asked which aircraft he was flying that day, he instantly recalled,

"It was airframe Bravo, which was actually christened 'Delphi' after the ancient holy city. It was very hard for the crew to comprehend, at that time Olympic employed nearly 8,500 staff and had been in existence since 1957. Aristotle Onassis, the shipping magnate had bought the airline from the Greek government back in 1956.
In those days it was called T.A.E the rather less attractive abbreviation for 'Technical and Aeronautical Exploitations'. In a re- branding exercise the company was renamed Olympic Airways reflecting Onassis's admiration for ancient Greece. The airline flew under the Olympic Airways name for over 40 years, before a renaming in 2003 produced the modernised 'Airlines' instead of 'Airways'. The glory years of the 1960s were synonymous with an airline operating in lavish style.
Captain Andreas said,
"On first radio contact with Athens Radar we requested to carry out a flypast for the honour of the airline. This would be the final time Olympic Airlines would arrive into Eleftherios Venizelos International Airport with long-haul passengers. We had flown all through the night across the Atlantic and then

the length of Europe to bring them home. The Air Traffic Controllers were keen to grant our request for the low-flypast and offered us runway 03 left." Captain Andreas was the pilot handling the approach and the First Officer was monitoring pilot. He said of the low pass, "I remember the Co-Pilot saying, "that's low enough now" at 80 FEET on the Radio Altimeter.

I remember the Co-Pilot saying, "That's low enough now..."

The passengers were cheering and clapping loudly as we flew down 03L past the control tower. We could hear them cheering and applauding from inside the flightdeck, it was an emotional feeling I can tell you." Captain Andreas continued to explain,

"It was necessary to pull the Circuit Breakers for the Ground Proximity Warning System to prevent a nuisance warning as we flew so close to the ground at such a low altitude. Of course, this was something which we would never have done in normal line operations, but this flight was different... It was the last one."

On the subject of configuration for the low-pass,

"We had discussed it as a crew before we started our final descent towards destination. We elected to fly the aircraft at 160 knots at Flap 2 setting with the landing gear retracted. This would give good stability and full control of the aircraft. It would also allow us to show the airframe for a significant period of time to the ATC Tower. They were very impressed and grateful for the fly-past."

Then it was time to land and taxi in for the final time. He described the emotions of the crew and passengers,

."..well *everybody* was crying. The crew members were in tears and the passengers also. We had some Olympic Airlines staff coming home from the outstations and they too were sobbing. Everyone was hugging each other.

I remember that all the cabin crew kept their sunglasses on, so you couldn't see their eyes. If you look at the TV news coverage of that time, when they were being interviewed they wore sunglasses even inside the terminal!"

It was an historic occasion, the airline had been flying for 42 years with great success and this was a very sad day for us all... It had been an honour to serve with the company and to fly the last flight."

End of an era - the crew of the last OA long-haul flight home to Athens.

Olympic Airlines had an interesting history. It had been running since 1957 having been bought by the shipping magnate Aristotle Onassis in 1956. He renamed it Olympic Airways and wanted to use the five colourful Olympic Rings as the logo for the company. The story goes that the Olympic

Committee organisation objected – they had the design patented symbolising the five continents.

Onassis simply instructed his marketing, publicity department to add a sixth ring – a gold one of course to represent the airline. At first glance you see only five rings, but then you see the sixth one.

Olympic Airways - six rings on the logo.

4 BIG TEAM

By Capt James McBride

I saw a hand waving at me, one of my colleagues.

It was while I was doing the walkround check on an airliner that it hit me. Another aircraft in the colours of our company, swung into the adjacent stand, dowsed the taxi light and continued towards the parking position at the gate. As it did so, I looked to the flightdeck windows and saw a hand waving at me, one of my colleagues. It felt good to be part of the team. I thought; *...Big Team...*

You see the aircraft I was operating (a Boeing 737) is quite small in industry terms. Known as a narrow-body or a single-aisle jet, the -800 variant carries less than 200 passengers in all economy configuration. In our world it is often the wide-bodied, long-haul aircraft which attract most kudos. That being said, the numbers tell a different story. If you total up the number of individual passengers carried each flying day, then... *'Respect'* to the short-haul airline crew.

Most companies require their pilots to operate 4 sectors per day in the short-haul flying programme which means nearly 800 passengers are carried in a flying duty period. There are not many wide-bodies which have that sort of capacity and some days we are rostered to fly 6 sectors... So, over 1000 pax in a day? A BIG job then – yes, and a Big Team required to carry out the mission. Seriously the fleet of aircraft is massive and the numbers of crew employed to operate on the line are prodigious. The training, management and supervision of so many individuals is a mammoth task for a company.

A lot depends upon the quality of the people themselves and their commitment to their working colleagues. Every working day we have to don a uniform and behave in a proscribed manner such that our co- workers respect us and see that we also respect them. If this does not happen, then the system doesn't work. In the old days it was referred to as 'courtesy and common sense' these days it has been made into a science called CRM.

For example, I was trained that we stay together as a crew, pretty much always. We report for duty as a crew, brief as a crew, operate as a crew then cease-work... as a crew. We are... The Team. Some time ago, I recall we reported for a straightforward two sector day. It was an out and back Alicante flight from an airport in Germany. Unfortunately when we arrived onboard the aircraft it was found to be unserviceable due to a technical malfunction. The fault was found by the First Officer (*PF) when he carried out the preflight system tests. The aircraft failed the 'Fire Test', which was a no-go item according to the MEL. We called

Maintenance Control in airline HQ and they said they would send an engineer to rectify the problem. Meanwhile we had already boarded the passengers and now we had to explain to them the reason for the delay to our departure which was essentially indefinite. The Engineer was coming from another Maintenance Base and would take at least 3 hours to arrive, hmm... what to do? I called the dispatcher over.

"Okay, put the passengers back in the terminal – they will be more comfortable there, I will speak to them now."

I explained the problem to the Pax* on the PA from the forward crew station in the cabin and assured them we would get them to their Spanish destination as soon as possible. Always when you send the passengers back to the terminal it is important to send them with all of their cabin baggage and belongings because you cannot predict what is going to happen next. It was just aswell that I did, because Ops then informed us we would be swapping aircraft and picking up the next inbound machine for our Alicante mission – this was due into the airport in one hour and twenty minutes.

This was an improvement which I was pleased about, however our First Officer saw things rather differently, he said,

"Captain, I've worked out that we will be back too late for the midnight local curfew here in Germany, so the return flight will probably divert and I can't take that risk..." We discussed the projected timings of the flights and I could see it would be close, although Ops had assured that we would have a high-speed flightplan each way etc.

The First Officer then advised that he had recently resigned from the company and only had five working days left to do. Clearly he was not going to be the most motivated employee, but regardless I would have expected him to have some sort of moral sense of duty towards his fellow crew members – it was not to be.

After the passengers had disembarked, we waited for a crewbus to take us all back to the terminal – there we would reprint flightplans and the briefing package for the new aircraft and monitor the progress of the inbound aircraft we would be taking over. The FO meanwhile had arranged his own ride to the terminal with one of the dispatchers and simply said to the rest of us (HIS CREW), "I'll see you in the crewroom, I'm going to talk to Crewing now and tell them I can't do it... I just can't take the risk of a diversion y'see...?" - as if that would explain and justify his actions to the rest of us?! Duh!!?

Fast forward 30 minutes later and we finally arrived back at the crewroom (it took a while to get the crewbus). As we were walking in, he was walking out. I said to him,

"Is that it? You're off then...?" To which he replied,

"Well! I couldn't get through on the phone to Crewing AS USUAL..." he raised his eyes to heaven at this point in dramatic fashion then continued, *"so I'll send them an email."*

And that was it, the last we saw of him. He left the airline (our Big Team) a few days later and I don't what happened after that. And what of us...? His abandoned co-workers...? Well I reprinted all the flightplans myself, kept the crew informed and motivated to do a good job, then talked to the handling agents about the passengers.

Also I called Crewing, hung on for a few minutes, but no answer, clearly they were busy. I still had to inform somebody soonest however that in addition to everything else, we were now a First Officer short. I called Ops. The lady Ops Controller on the phone picked up within 4 rings – I told her briefly what had occurred.

Understandably she was incredulous,

"He said WHAT!!??" was one of her phrases I recall. I tried to placate her with,

.".. well he was very polite about it, he shook our hands as he said farewell and wished us good luck! Oh, and he did say he would send Crewing an email..." on the other end of the phone I could sense the grinding of teeth. I continued with

"Ahh... anyway, I thought you ought to know as soon as possible in the light of our dilemma with the curfew." I smiled at the ridiculousness of the situation. Now, after all that had happened it was a distinct possibility that we could run out of time if they couldn't get a replacement FO. Maybe even both sectors would be cancelled, which would mean the holiday plans of nearly 400 customers would be severely disrupted, all because one employee was not prepared to be flexible.

At that moment the other phone in the Crewroom rang and they called my name – it was the handling agent to inform us that our new aircraft was taxiing in. I said to Ops,

"So... I'll leave it with you, we'll continue to get on with it and load the Pax ready for a departure when another FO joins us." I rounded up the cabin crew and we got on another crewbus to go out to the ramp again. One of the Juniors, a steward who had only been with the Big Team for about 12 weeks was curious about where the FO had gone and why. When I told him he said,

"Well I don't want to divert either..." implying that maybe it was now a matter of personal choice as to whether we flew to Alicante or not. I replied that in my case I still needed the job and perhaps he did too, so we would just have to suck it up and get on with it. The other cabin crew looked a bit down in the dumps too. You could understand this when they had just witnessed one of the pilots bailing out.

I tried to cheer them up with,

"Remember these passengers will be tired, hungry and thirsty. Just keep smiling, stay cheerful – you will sell-out on the bar service and make loads of commission!" On the ramp we were there as our new transport of delight arrived on stand, we waited for the Pax to get off then we boarded for our flight. A new face appeared as I was completing the external check of the jet.

"Hello, I'm your new First Officer!" he said brightly. "I've just been called by crewing to swap from another flight." Another member of the Big Team and I was delighted to see him.

It was the fastest Alicante flight we had ever done and the cabin crew did indeed sell virtually everything they had to offer. They were very pleased with their commission. The handling agents, Ops, refuellers and crew all worked hard to achieve the mission. We landed with just 10 minutes to spare before our base airport closed for the night.

A good result. I couldn't help wondering about the thought process of our original co-pilot however. With an attitude like that to 'duty' and to his fellow colleagues, I bet he will find that, yes it is a Big Team... but a small industry!

*PF – Pilot Flying, the one who actually handles the controls

*PAX – Industry abbreviation for passengers

It's a Big Team.... but a small industry.

5 BEING SCARED

By Capt Nick Francis

"Have you ever been scared?" How many times have we as airline pilots heard that question from a passenger visiting the flight deck.

"Oh yes, one of the girls told me she was late one month.... That was pretty scary!" I would wink and use this somewhat politically incorrect reply as a way of bouncing the mood back to one of jollity and enjoyment for the passengers. For whilst the questioner might be seeking some vivid tale of steely eyed determination to counter flaming engines or thunderstorms from hell, the reality is that while pilots might recall such episodes, the task of explaining the technical background would usually see the visitor become bored and lose interest.

Having answered this and many other predictable questions I would always leave my guests reassured as to their safety and wish them a happy flight. However on more than one occasion I would twist back into my seat and settle my eyes on some perfect scene of tranquil sky and calm seas some seven miles beneath me, and that question would echo in my mind.... *"Have you ever been scared?"*.... "Oh yes Madam, I've been very scared indeed."

For how often would that question take me back to the day when the view from the flightdeck was not one of beauty, but of a steel grey sky lashing rain across a foaming sea with fuel gauges that read zero. It was August 1978. I was one of several 'twenty something's' who were carving their careers as single crew pilots working for a charter airline out of Aberdeen.

Our work supporting the oil industry always took us to the Shetlands, Orkneys and Scandinavia in the foulest weather that nature could provide and frankly summer didn't seem to differ from winter except that the rain was warmer and the airframe icing started higher up.

As many will remember, the mid to late seventies were the 'Klondike' days of aviation around the North Sea. Rigs had to be fed their crews by helicopter operators from the Shetlands. Time was money, and in many cases seriously big money. Always there was commercial pressure to get the job done. It was quite normal to get a carpeting in the Chief Pilots office if you diverted when competitor airlines made it in, especially on a bad weather day.

The threat of having your career restricted or even being fired were ever present sticks with which we were beaten, and in reality whilst the rules were obeyed in spirit, they sometimes weren't in practice. Thus we learnt quickly about surviving in a world caught between tough weather and tough management.

On this particular day, at around six in the morning, five of us were assembled in the briefing office cursing the fact that the remains of a tropical hurricane had arrived in the UK and we were facing some appalling weather from Aberdeen to Sumburgh. Heavy rain and high winds were going to give us an unusually unpleasant day. Also, apart from the aircraft of my own company we would face congestion from the other operators as we all sought to plough our way north at the same time. There would no doubt be someone carpet dancing in the Chief Pilots office tonight.

Now our company had been the first to take the Embraer Bandeirante into service at Aberdeen and whilst it was a lovely aircraft to fly, being faster and more comfortable than the Twin Otters against which it was pitted, it did burn a lot of fuel. The trouble was that it burnt too much fuel to do what was asked of it which was to carry twenty hefty male passengers, their bags and a stewardess to the Shetlands. Whilst the flight manuals said it burnt a certain amount per hour, reality proved different.

So we found ourselves often taking the unpleasant option of carrying a small amount of fuel in the tanks that never appeared on paper to restore to us what was a sensible amount to cope with the air traffic congestion and the weather. We took comfort from the fact that in its native Brazil the aircraft operated to weights much higher than we did. Today would certainly mean carrying ghost fuel!

Sadly it was this morning that a member of management came into the crew room and stated that the carriage of such fuel must cease and that loadsheets and uplifts would be checked to ensure we were not overweight. *Oh wonderful* I thought. *A great day to go legal!* I considered arguing the point, but as I knew it would be futile I accepted like the others that we would operate within our permitted takeoff weight.

Boarding was normal as I stared at the fuel gauges and hoped that the flight was going to be routine. I didn't hold out much hope as I noted that the holding point queue stretched back to the apron as a variety of types from Heron through Viscount to Sikorsky S61 slowly and somewhat wearily trundled their way ever nearer to runway 17 and their chance for freedom to escape the lashing rain.

It was my turn, after a twenty-minute wait I lined up, still no airways clearance, but when it came I knew things must be busy. "Zulu Kilo, clear take off to climb on runway heading to flight level one one zero." What!!? It was unheard of to get a straight climb to that sort of level and it presented me with immediate problems. The first was that it would take me around twenty minutes to get that high at maximum takeoff weight, the second was that it was in the wrong direction, and finally the Bandeirante wasn't pressurised to climb above ten. However as we occasionally crept above that height for brief periods of traffic congestion I figured that negotiating on the way up would sort the problem out.

As the wipers pounded the rain left and right across my windscreen Aberdeen disappeared beneath me before the gear was fully up. It was going to be a flight without any view today. I called approach and was told to maintain my southerly heading and continue climb to eleven thousand. Protestations about lack of pressurisation brought little sympathy; the advisory route was just too busy.

On levelling off I was finally given a clearance to turn north and begin my journey to Sumburgh. By this time I was already aware of being woefully down on fuel, but a check on the ground speed showed that I was getting a 70-knot tailwind, way above the forecast and for the sort of levels we normally flew at it was a phenomenal speed. I cheered myself up with the thought that at least I would be making fuel and could see how the situation looked further up the route. I could always go to Wick or Kirkwall if things got bad.

It was normal at that time to speak directly to the met office at Wick when the airport was closed. Using this facility we could get the latest update on what was often very changeable weather, today though, no reply. I checked Kirkwall and found that although the weather was just about usable, landing there would involve either a twenty-five knot tailwind or some very nifty low level circling for the opposite runway. It was not looking good, but at least my fuel state was starting to cheer me up and I figured that I might be back on the line by the next waypoint. At this point I also got clearance down to nine thousand. Things were not too bad after all. It was time for breakfast.

Kim was the flight attendant that day. She had handed out the meals and came up with a much-needed coffee. She was having an easy time. Outbound in the morning the oil rig workers were more interested in the morning papers than her. The return passengers would not have seen a good-looking woman for nine weeks. She would not have such an easy flight back!

At the next waypoint I realised that I was still down on fuel, but things were getting better and I still had my good tail wind. Sadly at this point on the route I had no means of checking my ground speed accurately so whilst I reckoned I was still making time I could not be sure. The next waypoint beckoned, and after all Kirkwall was just about usable wasn't it?

Then my world changed. It was the Scottish controller 'All traffic, be advised that the approach radar at Sumburgh has just failed. Be prepared for extensive holding *Oh shit! Chief Pilot's carpet here I come!* I didn't waste time. I immediately asked for a diversion to Kirkwall and turned to the west. I was told to call them but found I was just out of range to get any reply.

Never mind they'll be there in a minute. Eventually they replied. I told them I was inbound and asked for the weather. The reply was not what I needed, gale force winds in driving rain and a cloud base down to 100 feet above ground. That meant the wind was totally out for an instrument approach and the cloud base too low for circling even by North Sea piloting standards! I asked for the updated Wick weather.

They tried to get it but to no avail. This was not good. I was now somewhere over the North Sea at nine thousand feet with nowhere to go. There was no point in going north to Sumburgh, I knew I didn't have the fuel and at that time they relied on radar approaches to get you in on the worst days. There was no ILS. I could hope for a short reprieve in the weather at Kirkwall but God help me if I didn't get it.

I could head for Wick, which didn't open for another hour, but it only offered a VOR approach into unknown weather. Or I could do what I did, which was to state that I was turning for the north coast and RAF Lossiemouth. I had no idea if I would make it, or even to dry land but at least they had ILS and radar, and the thought of having my position known was a little more comforting if things were going to get as serious as they looked.

What's the saying, "cheer up, things could be worse"? Right now I was heading into as worse as it gets. Kirkwall changed me back to Scottish control and I declared my intentions. Funnily enough at this point things began to become a little surreal. I was asked if I wanted to declare an emergency, somehow I only heard myself say "yes" rather than think about it.

I was told to squawk a different code, and it was only after setting it that it dawned on me that it was the 'Mayday' code. I was changed to 121.5 the emergency frequency, and it felt just like changing to any other station. Was I losing touch? I don't know, but I knew by know that I was going to need every bit of help available and standing on pride would be fatal.

I slid the power levers back to the long-range cruise setting and ran around the local radio aids to try and get a groundspeed. Nothing! Everything was out of range. I had descended to six thousand to get out of the wind and found that I only had voice communication. It would be anyone's guess how long it would be for me to receive anything from the north coast. I had no idea if that phenomenal wind that had helped me get so far north so quickly was about to become my executioner and see me run out of fuel before landing.

It was at this point that for the first time in my life I became really scared. It wasn't panic. Pilots are not allowed to panic. No, this was real fear, not the 'fight or flee' rush of adrenaline that you get from a sudden event. Really being scared is knowing that you are probably about to die, but not until you've finished your job. I sat and contemplated the fact that we would most likely ditch just north of Lossiemouth and hopefully be found by rescue helicopter quickly, after all at least they would see us go down on radar, but so what.

There were life jackets under the seats, but no survival equipment. Everyone was in heavy clothing that would weigh them down, and in the water temperatures that existed in the North Sea we were told that the only difference between summer and winter was that in summer you lived for twenty minutes instead of ten... I was in shirtsleeves.

Real fear comes when you have time to think. Having an event like this unfold slowly is infinitely worse than a sudden occurrence. You have time to analyse. Time to curse yourself for being so stupid, time to imagine what may come, time to think of those you will leave behind. Was it not a pleasant experience to drown? So I'd heard, and I found myself hoping it was. Would I get the Bandeirante into the water intact? What was that about ditching? Was it along or across the swell? Along seemed right, but then what about the wind...?

I sat and contemplated what it would be like to see the water rising through the aircraft. I couldn't swim at the time and I realised that however well I ditched the aircraft I was probably going to die either in or after the event. How could I have been so stupid as to get in this mess! I'd been a flying instructor. Hadn't I lectured enough students about the tendency to carry on regardless? Had the threat of commercial pressure made me go that bit further than I should? I should have gone back into Aberdeen ages ago when I still had the fuel. I should have... I should have... and so it went on in my mind.

But then a thought entered my head and started to repeat itself, over and over it went, almost becoming a mantra... *No God, not today! You're not having me today or anytime soon. You've tried often enough in the past, but not today!...* I hung on to that thought, and anytime the fear of what was to come rose up in a surge from my stomach I would shout it aloud in my mind... *No God. Not today!*

I called Kim up front and explained what was happening. I said that I would know well in advance if we were to ditch, and for the time being just keep calm and give some thought to her ditching drills, there would be plenty of time for mental rehearsal. I told the passengers that we would be landing at Lossiemouth, but at present I didn't know when. I hadn't taken much interest in them since taking off, but as Kim turned away I spotted something different.

A man sitting on the front row, wearing a grey suit and carrying a black leather brief case. Not your average oily at all. He sat bolt upright cradling the brief case to his chest, almost appearing fearful of having it stolen. Whilst others read their morning papers his eyes were fixed straight ahead. Perhaps he had sensed our danger, perhaps he knew of our predicament. I doubted it, probably just fearful of flying... and with very good cause today!

Scottish Control came up and asked me to speak to Lossiemouth who by now were beginning to pick up some weak radar returns. The subject was what I wanted to do about ditching. Preferably not do it was answer number one, but they wanted to know if they should scramble the helicopters now or wait for me to declare my intentions, in the meantime they would ready a Nimrod to find me and lay a path of smoke flares so that I could have wind and sea level indications at touchdown. It would also lay life rafts alongside the flares for those of us lucky enough to get out. This was not a conversation I was enjoying... *No God. Not today!*

Finally the DME from Lossiemouth burst into life. 74 miles. At last, at least I could get a ground speed. Cruise was 250 knots, ground speed 180, Jesus! Still 70 knots of wind even down here at six zero *('I'll give you this God, you're really trying hard today'.)* Time to Lossiemouth was 30 minutes allowing for the approach. I looked at the fuel gauges and saw that they read a total of 130 kg. Consumption was 260 an hour.... endurance 30 mins. Maybe a touch longer when we throttle back for the ILS... *No God. Not today!*

We flew on. The windows still as steel grey as they had been since take off. The DME clicked its digits downwards. I sat in silent contemplation with my mantra. Kim brought a coffee, but somehow I wasn't in need of it. Ground speed increased a little and I eased the fuel flows back wondering if the PT6's were as hot tempered in their tail pipes as I was for getting myself into this mess. Lossiemouth finally picked us up on radar and gave me a heading to steer. I had made a lucky guess, it was only ten degrees right.

It was decision time. I knew it was coming. After all I knew that eventually I would have to decide if I was going to make a controlled descent into the sea or carry on for a landing. Have I not been sitting here for over an hour knowing that sooner or later I would have to decide my own fate and that of the other twenty-one people on this plane? Lossie ask me for my intentions. They have to scramble the Nimrod and the helicopters if we are to meet. I look at the fuel gauges, tap my calculator, put in my wish factor and realise that it's an even bet. We might make it, there again we might run out of fuel on final approach. If I ditch, then we will certainly either drown or die of exposure, to carry on will at least give us a chance. No contest.....we carry on... *No God. Not today!*

We crawl ever closer and the ground speed starts to increase as the controller guides us out of the wind and onto the localiser, 190, 200, 210 knots. Power is eased back with each gain and fuel flow reduced accordingly. The latest weather is to hand, am I ready to copy? Oh good! A twenty five knot crosswind, 1000 meters in driving rain, and a cloud base of 200' *(Oh God! you really are going for it today).*

Ten miles out and the fuel gauges are now on zero but we are on the ILS. The turbulence is horrendous, and the ILS needles seem allergic to meeting each other in the centre of the instrument but I find myself and my mantra carrying us down the glideslope. Finally we break cloud and I spot the sea out of my left window. It is white, not grey, a total furious mixture of churning mayhem, wilder than I have ever seen water before. The rain lashes the windscreen and the approach lights come in sight. I kick off the drift, drop the left wing as low as I dare without scraping the tip and we touch on the left main gear. We are down... *No God. I told you, NOT TODAY!*

As I clear the runway the tower ask for my endurance. I look at the gauges on their stops and the left one flicks a little in the turn. 'Five minutes' I reply, but I doubt it.

On the ramp I pull the fuel levers to cutoff and pause for a moment as engine noise gives way to wind and rain. Oblivious to the noise of the departing passengers I contemplate the raindrops rolling down the windows. Only one thought seems to be in my head now, and I quietly whisper it to myself... *Thank you God, Another time I guess.*

After lunch and a check on the weather we set off back to Sumburgh. The radar had been repaired and the flight was uneventful. Kim fought off the best advances of twenty sex starved 'oillies' on the way home and we both left the office after what had been one hell of a day. As we parted at the door I asked her a question. "Hey Kim, who was that guy with the briefcase on the front row?"

"Oh, he was a marine accident investigator on his way to a rig to investigate a fatality. A couple of guys got killed diving yesterday. I guess he was going to find out what happened." I gave her an ironic smile. "Then I guess he'll never know how close he came to asking them personally,

Goodnight Kim."

So did I get a carpeting? Not really. I stood on the Chief Pilot's carpet alright, but I think the management were so glad to see me alive I thought they were going to hug me. Did they ever raise the subject of carrying company minimum fuel again? No.

Nowadays I still fly a twin over water, but now it's the oceans of the world in a 767. Am I ever tempted to take company minimum fuel?... Well not often.

But I guess you can understand why.

Editor's note – Airline passenger visits to the flightdeck in flight used to be a common occurrence in the days before September 11th, 2001. After that date, the airline industry locked the flightdeck doors forever.

Cruising mid-Atlantic in a Boeing 767.

6 A GOOD SHIP

By Capt James McBride

I was doing the pre-flight walkround on an airliner last week and once again it struck me how many parallels there are between the aviation industry and maritime. For some reason I am always drawn back to the descriptions of Captains in Nelson's Navy making external inspections of their ships sometimes from a small boat being rowed by their sailors.

As my 'Command' is on dry land, I have no trouble in moving myself round the outside to visually inspect the machine. Naturally in our safety conscious age I am unable to do this without a bright yellow Hi- Vis waistcoat and ear protection.

While I am doing 'the walk-round' as it is known in the trade, I am making a very careful inspection of the condition of the airframe. I am conscious that this is the last time the aircraft will be inspected before we fly, so in my mind it is the absolute last chance to see something which might be amiss. Of course, the engineering team has already done a thorough check themselves – they will have checked the tyre pressures, engine oil levels, hydraulic system contents and probably even had the cowlings open on the engines.

Coincidentally I met with the engineer as I arrived at the aircraft with the crew this morning and he had completed a '50 hour check' overnight, which meant that the team had been even more thorough than usual. His final words to me, after describing a couple of minor defects were, "Overall... she's a good ship." As simple as that - a direct reference to the days of maritime commercial transport. I smiled when he said it, but I don't think he will have guessed why. *So many links to the past, even when we are racing ahead with aircraft design into the future* – I mused.

As if to illustrate my thoughts, less than a minute later, two Airbus A380s and a Boeing 787 Dreamliner taxied past only metres away from our parking position. The most modern airliners operating in the world regularly fly in and out of this airport. But even these brand-new machines are steered on the ground by their pilots using what is referred to as a 'tiller' and at the back of the fuselage is the rudder...

Continuing my walk-round, I inspect carefully the state of the landing gear, then stick my head as far as I can into the engines. In comparison with a sailing vessel, I would say that the wings of our aeroplane equate to the masts, while the engines are the sails. Both provide the motive power to move the vessel. Of course when it comes to talking about the fuselage of an airliner, more often than not you will hear pilots and aviation engineers discussing "the hull", exactly as you would with a ship. I look at the hull carefully, noting any marks which might be evidence of ground damage.

I am confident in the knowledge that any previous damage to the aircraft should be noted in the 'dent and buckle charts' which are kept in the flight-deck. Note the use of the term 'deck' of course. In fact in some airlines of the recent past (Dan-Air in the UK was one of them) they used to talk about inflight *pitch attitude* as 'Deck Angle' – no seriously, I'm not making that up, it is true. A nautical expression if ever there was one. Then we have more references to decks. After the Flightdeck there is the Upper Deck and Main Deck on the larger airliners. Not only decks, but bulkheads too are named parts of the aircraft structure – a continuation of maritime tradition.

In keeping with naval tradition, it is logical that the aircraft Commander should be referred to as 'the Captain' although; perhaps less obvious is the airline rank of 'First Officer'. This is likely an amalgamation of 'Chief Officer' and 'First Mate' which are the two accepted names for the second-in-command onboard ship. In many traditional airlines the Senior Cabin Crew Member would be referred to as 'the Purser' – just like being afloat. The parallels continue with the official documentation that surrounds the operation of air-liners.

In the flightdeck is usually stored an important legal document which records all details of previous flights with the airframe. The Aircraft Technical-Log (ATL) is ubiquitously known as "The Techlog." Not only previous operational data, flight times, airports, numbers of landings and takeoffs etc., but also most importantly the engineering/maintenance information are included therein. 'Log' is actually a very old naval term indeed. The origin can be found in sailing ships from hundreds of years ago when a real wooden log; tied to a long cord with knots in it was dropped off the stern of the vessel and then the length of string was timed with a 30 second sandglass as it ran out.

The longer the string, when the time was called (more knots) the higher the speed of course. In such a way a ship's crew could record their speed and distance travelled. Of course there were plenty of errors in the method as it only measured the ship's movement relative to the sea. To be more accurate one would have to make allowance for current, tidal effects etc.

In a similar way an aircraft's speed is usually measured in Knots - nautical miles per hour. There are many different ways of considering/ calculating speed in Knots, Indicated Air Speed, True Air Speed, Rectified Air Speed (also known as Calibrated Air Speed), but the one thing is common – it's all in Knots. The reason for the *Nautical* MPH is that it is more logical.

One nautical mile is equivalent to one minute of latitude on the earth's surface (or one sixtieth of a degree of latitude). That being said, the world is not a perfect place, especially in its shape – the sphere is not spherical. Which means that the length of a minute of latitude is 1861 metres at the poles and 1843 at the equator. In fact the NM is now standardised as an SI derived unit*, rounded to an even number of metres 1852**.

I remember very clearly when I was a new First Officer on the Boeing 757 some years ago now, sat in the flightdeck listening to the captain as he gave a preflight summary of the technical status of the aircraft. He would usually start with where the plane had arrived from, and then describe any technical snags which were being carried forward, finishing off with the Autoland status... *"...and she's a Cat 3 Ship."*

This final piece of information was very reassuring because it meant that even in the poorest met visibility at destination, we had all the functions of triple ILS, triple autopilot and triple autoland – we would get in! The thing which struck me every time though, was not the sophisticated, modern technical capability we had at our disposal, but the simple reference to "...the Ship." There it was, in plain terms.

The Master and Commander referring to his 'Command'. Not only that but, he also acknowledged that the vessel was female – ships are always she. At the time I mused - *how quaint?* But here we are 30 something years later and I find myself saying the exact same thing to my much younger colleagues... *"...she's a Cat 3 ship!"*

Every flight is a voyage and at the end of the voyage our ship is sometimes met by a 'follow-me' car and marshallers to guide us to the final parking position. As we taxi slowly onto stand with their guidance, it is redolent of the way in which a maritime 'Pilot' in a foreign port guides the captain into harbour.

As we come onto stand and I apply the parking brake for the last flight of the day and after confirming AC electrics are established, I shutdown the engines. Sometimes I get a strange look from the FO when he hears me murmur... *"Finished with Engines."* If he's reading this now, he will know why I say it.

** SI = The International System of units. ** The international Nautical Mile was officially defined as 1852m at the First International Extraordinary Hydrographic Conference in Monaco in 1929.*

...this is the last time the aircraft will be inspected before we fly.

7 MEDICAL DIVERSION TO CORFU

By SFO Jon Mytom-Hart

B737 Airliner – from Cruise Altitude to landing in 22 minutes

We were flying a scheduled service from Bologna back to Athens at 37,000ft with a flight time of approximately two hours. An hour or so into the flight we were just checking in with the northern Greek controllers when we heard the double chime from the cabin signifying that there was a problem. I was handed the radios by the captain as I was Pilot Flying and he spoke with the Cabin Supervisor on the intercom.

Meantime I was admiring the view on a crystal clear day over the Ionian Sea. I could see the island of Corfu out to my right. My attention was brought back to the flight deck when the captain said

"Right Jon, we have a problem, there is a 30yr old woman who has suffered an Epileptic fit and as a result her heart has stopped. There is a doctor onboard who is now treating her, but we need to divert and get medical attention for her as soon as possible. What do you think our options are?"

After expanding the scale on our Nav Display we saw that both Athens and Thessaloniki were too far away and would take approximately an hour or more to get on the ground. The other destinations directly ahead were not familiar with either of us as a crew or the company. Corfu however, was just at our 5 o'clock within 50 miles and as it happens, was not only a destination but a company base. I suggested we divert to Corfu which was quickly agreed by my superior in the lefthand seat.

The captain made a brief announcement to the passengers and crew over the PA system and we put the seatbelt sign on. He then came back onto the radio frequency with me, made the PAN call and the Greek controller gave clear instructions to turn left direct to Kerkira VOR (KRK). We were cleared for descent to Flight Level 100.

We acknowledged the instructions and requested medical services upon arrival. At this point we were both under the impression our passenger was still without a heartbeat and commenced the emergency descent. This was a manoeuvre I had only done in the simulator up to this point. In short we had the aircraft descending at maximum speed allowed with the speed brakes in the flight detent position to allow us to get down to the cleared level as quickly as possible.

When we were approaching FL150 we were handed over to the ATC Tower in Kerkira who then re- cleared us to take up the hold over KRK VOR, continue our descent to altitude 3,500ft and for us to advise when we were ready to commence the approach. We arrived at the VOR beacon just below FL100 and I slowed the jet down to take up the hold and give us time to set up the aircraft for what was going to be the Non-Precision VOR approach to Runway 35 in Corfu.

If you have done this approach on a CAVOK day without the pressure of an emergency, it is truly a spectacular procedure to land an aircraft on an island. The Captain was busy setting up all the navigation aids and computer routing needed to be able to commence the procedure when we got a call from the cabin once more which stated,

"The woman has been revived with CPR by the doctor, but still needs urgent medical attention" When we received this message it must be said that there was a huge sigh of relief, but our focus was still to land as soon as possible. I handed over control to my colleague and briefed him on what to expect and what he had set in the aircraft computers.

Also we cross-checked this with the Navigation Database on our iPads. By this point we were on the outbound leg of our first hold and approaching 5,500ft in the descent. We received a call from the Tower asking if we would need another lap around the hold before the approach. The captain looked at me and said,

"What do you think Jon? We are at a good height with a reasonable speed," I replied with,

"Yes I think we're good to go, but let's just double check the Airfield brief and our minimums before committing and missing something vital," he smiled and replied that this was a good idea as it's always easy to rush and miss something that could make things very difficult.

Thirty seconds later we were both comfortable and announced we were exiting the hold and commencing the approach. In response, the Tower cleared us straight to finals. The cabin was secure and all was now seemingly like a regular stage of flight (whilst trying to ignore the situation in the cabin and concentrate in getting this correct on the first attempt).

We were on final approach now and nearly 500ft when I saw out of the corner of my eye what I thought looked like a bird. I really hoped this wouldn't be the icing on the cake. We descended another 200ft and then another bird that came into view very quickly and met a quick death on the nose of our jet followed by yet another flashing past my right-hand window. I had briefed the Go-Around procedure prior to leaving the hold and I was now playing this through once again in my head, only to be reassured by the captain calmly saying

"Continue." The Go-Around would have been challenging with high terrain to our left, in-front and right so we didn't need this extra stress. The landing was normal, although after touchdown we vacated the runway promptly. Tower gave us instructions to taxi to our parking stand and we prepared the aircraft to be shut down and doors opened as quickly as possible.

Whilst running through my checks (omitting the flaps as we would most certainly need to inspect the aircraft for the remains of the birds that met a high impact death). The captain jumped on the radio and said,

"ERRR Tower where is the ambulance that we requested?" To which the reply was,

"Yes we have contacted them and someone will be there soon..." We looked at each other and all I could say was,

"If there is one thing about landing on an island it is that things never happen quickly."

The engines were shut down and we released the cabin crew to aid our passenger in vacating the aircraft. Only then when we came out from the cockpit were we finally able to see the situation with our own eyes. Not only had the woman recovered, she was standing up, apologising and begging to continue to Athens!

This was obviously not an option as we had made every effort to get her on the ground as fast as humanly possible at the inconvenience of the rest of the passengers. As you can imagine this caused a rather big upset within the cabin to which the crew was faced with calming down hot, angry and tired members of the public without much of an incentive to help them. I removed myself from the cabin to inspect the outside of the aircraft and report to the captain my findings while he communicated with the handling agent and also the medical staff.

Whilst walking round the jet I took time to reflect on our actions and as most people would, I wondered if we could have done anything better? It was then that I noticed the bird we hit on the nose and that nothing much was left. The other bird had struck the left outer cowling of the right engine, but there wasn't evidence of it going through the engine air intake so that was a relief. I distinctly remember banging my head on the fully-extended trailing edge flaps and cursing what was turning out to be a long day! I returned and told the captain what I had

seen and he said he would also check before calling engineers to inspect and sign us off to be able to continue to Athens.

Author, Senior FO Jon Mytom-Hart - now flying for Emirates.

8 DELIVERY FLIGHT

By Capt Steve Cross

We used to find these tiny bush strips in the middle of nowhere somehow.

I was working for the Flying Doctor Service in Western Australia, you know the distances are vast. I mean not for nothing is the middle of Oz known as "the GAFA" - that stands for the Great Australian Fuck All, by the way...

We had a fleet of light twin engined aircraft and used to operate to some pretty out-of-the-way places as you can imagine. Of course we didn't just fly the Doctors out to the patients, most of our work was flying the patients themselves, often when they needed urgent medical assistance.

Not all of our aircraft were fitted with autopilots, but fortunately the one I was flying was, on the day I recall being tasked to divert to pick up a patient, a young aboriginal woman who was pregnant.

When I first saw her she looked about 12 years old, but I guess she was around 17 and heavily pregnant. The flight back to base where the hospital was would take about 4 hours. All our aircraft were fitted with extra-long range fuel tanks.

Anyway, we loaded her aboard, (me and the other members of the tribe). They said farewell to her and I was surprised that none of them were going to accompany her for what looked like an imminent birth... We made her pretty comfy in the cabin, so I got into the cockpit ran through the checks and we took off in a cloud of dust from the outback strip.

I really don't know how we managed in those days before the advent of GPS. We used to find these tiny little bush strips in the middle of nowhere somehow. Naturally it helped if you had been there before, but you always had to find it the first time on your own.

We had only been an hour airborne, when I heard a noise from the cabin behind me and it was clear she was in pain. It was a loud groan and then another one and another. Oh Crikey! I thought this ain't good. I got straight on the radio to base ops, there was always a nurse on duty who could give us advice in flight, but I very rarely had to use it. Besides which I was on my own, what was I going to do?

Her voice came clear as a bell on the radio...

"How often are the contractions?"

Christ! Contractions!!?? Is THAT what they are?

."..well, it seems they're only a couple of minutes apart." She replied,

"Steve it looks like she's going to have the baby inflight, now is anybody with you?" At that moment I felt SO alone ya'know. In fact I WAS ALONE!

Anyway, the nurse continued with her advice and instructions,

"Put her on the stretcher, so she can lie down..." We carried a stretcher in the back of most of the aircraft, but she wouldn't be able to get up on to it without my help.

"Hang on love!" I said, "I'll come and help you on to the stretcher" and with that I took off my headset and slowly got out of my seat. I didn't want to knock the control column at all because the Autopilot might kick out.

In the same way, I double checked the throttle frictions – they were all screwed in tight as they would go. I felt like I was going to be a wing-walker... Well the girl was panting away and groaning aplenty as I got her settled on to the stretcher and then I went back into the cockpit, grabbed the mike and said,

"Okay she's on the stretcher, now what?"

My heart sank as the nurse replied with a question,

"Is she wearing any underwear?"

"How the fuck should I know!?" I shouted back. I didn't mean to shout, but I was in a bit of flap now mate I can tell you. I could rapidly see where this was going... The nurse though was cool as ice, she simply ignored my outburst and said,

"Steve, you will have to use the blunt-nosed shears to cut her knickers off at the side, do each side one at a time carefully..." I guess she didn't need to tell me to be careful, but my hands were shaking, I could feel them. And sweating, man was I sweating.

I checked carefully that the autopilot was holding the heading and altitude, double checked the engines. I mean if that autopilot dropped out while I was back there, that would have been BOTH of us having a baby!

You see I was very young at the time, maybe 22 years old. Mate, I tell yer, I'd never even seen pussy in daylight...!

So I followed the instructions I was given and we started the delivery. The head came first and I did what they said, I put my hand under the little chap's scalp as he came out and slowly, ever so gently drew the rest of his little body out.

He was crying, I was crying, mum was crying... All I could think was, *thank fuck the autopilot hasn't tripped out.*

Then I followed the rest of the nurse's instructions to me. I mean after cutting the poor girl's underwear off, it was a piece of cake to chop the scissors up the front of her thin cotton dress.

"... cut the dress all the way up the front so the baby can lie between her breasts. Clamp the umbilical cord twice with the clamps, DON'T CUT IT STEVE, but it must be clamped."

I placed the little babe on her chest and covered him up with a towel and he instantly stopped crying, can you believe that!? Amazin'.

And then all of this other stuff started coming out after the baby, and I just wished it would stop ya'know. So I got lots of towels and cloths and stuffed them down there which helped. Anyway, then I got back to the job of flying the aircraft - I managed to get blood all over the controls which wasn't pleasant, but it soon dried off. On the radio they suggested I divert to ____ which also had a hospital and they even had a Tarmac runway. Mate I did THE SMOOTHEST LANDING I've ever done in my entire life. It was like butter, followed by the gentlest braking in aviation history.

They were both fine. The medical team met the aircraft and took them away in the ambulance. Then the nurses did a quick inventory and restocked the supplies of dressings etc.

I needed a BEER! But do you know what? They wanted the ship back in base ASAP, so I had to takeoff again and fly another 2 hours to get her home. THEN I had a beer!

I never found out how they got on mother and child. I don't think she named him 'Steve', but it would have been nice to know they were alright. Of course the last I saw of them they were fine. She was young and healthy and the babe seemed strong, he had a good pair of lungs on him anyways!

Well that was the essence of The Flying Doctor Service, you never knew what you would be doing from one day to the next.

Of all the flights I did THAT was THE most memorable, it was a real Delivery Flight!

Steve Cross is now retired. After leaving the Flying Doctor Service he went on to have a full and successful career with Qantas Airlines flying all over the world.

9 IF IT'S NOT BOEING...

Interview with Brien Wygle by James McBride

From biplanes with fabric covered wings to wide-bodied jet airliners – he saw it all. Captain Brien Wygle's flying career spanned four decades in aviation. His first pilot course was in 1942, he went on to fly as a combat cargo pilot in Dakota C-47s in World War 2, serving in India near the Burmese border with Canadian squadrons.

In 1946 he flew from England flying Dakotas in Europe, and then returned to his native Canada to start his peacetime career as a professional aviator. Enrolling in the University of British Columbia he completed a degree in mechanical engineering and at the same time he kept flying. Now it was with the Vancouver based RCAF Auxiliary squadron where he flew T6 Harvards, P51 Mustangs and significantly one of the world's first ever Jet aircraft, the Vampire.

Employed by Boeing as a Test Pilot after he finished Uni, in 1967 he flew the first ever B737 and ran the test flight programme of the aircraft. Also he flew the first ever B747 two years later. He finally retired in 1990 as Vice President Flight Ops.

Q. After such a career in aviation, you must have many memorable flights. Are there some which come more easily to mind than others?

"One of my most challenging flights was during World War 2 when I was flying C47s, the military version of the Dakota over the hump in Burma. These were difficult missions for us all because the conditions were not easy. The monsoon weather, the tropical rainforest climate and then we had the very high terrain to contend with. Flying unpressurised airplanes over the Himalayas supplying troops at the front was no easy task.

Well, one day we arrived at our destination; just behind the front line and on the walk-round of the airplane prior to flying it back I saw there was an oil leak from the right engine. The oil seemed to be leaking from the propeller governor at the front of the engine, when we checked the oil level, it was nearly empty. I was amazed that the engine was still running at all, let alone providing power and oil pressure. Although those Dakota engines were pretty reliable.

So, I thought about the problem and discussed it with the crew. I said if we have nearly run out of oil, then we ought to stay here overnight and get an engineer to come up tomorrow to fix it.

At which point I went to speak with the senior army officer there and told him. I said we proposed to stay the night with the troops, at which point he pointed over to a pile of wreckage and said,

"Well you're welcome of course, but that's what remains of the last Dakota when they stayed overnight – you see the front line is very close and sometimes it moves in the period of a few hours. On that night, we were over-run and their plane was destroyed..."

Well I went back to my crew and I said,
"We're going!'
I explained to them what I'd been told and reasoned that, well the engine had been running when we landed...

When the other engine, THE GOOD ENGINE, decided to quit! Without any warning, just 'Bang' and there we were with no engines...

We took off again and headed for home base, I climbed as high as we could go so that we would have time to sort things out if/when we had a problem. The big thing for me was that the engine had been leaking from the constant speed, governor unit and I was concerned that we might not have enough oil to feather the prop fully. We decided to do that early before it stopped running completely. So now we were running on one engine, but we had plenty of altitude and we were halfway home with the bad engine feathered (giving minimum drag).

That was when the other engine, THE GOOD ENGINE, decided to quit! Without any warning, just 'Bang' and there we were with no engines...

Just a glider! Looking for a field to land in. We looked down below and sure enough there was the little airfield we needed, we only had the one shot at it and the runway was short, but we made it okay. Years later at Boeing when we were discussing ETOPS* and the reliability of modern engines, I could add to the debate my experience, but really that was before the jet engine. Jet engines did become extremely reliable which is how ETOPS came about."

Q. And you carried on flying after the war ended, many pilots didn't?

"Well after the war was over, I wanted to keep flying so I went to the University of British Columbia, Canada and studied engineering. While I was there I joined the RCAF Reserves, it was a tiny squadron – everything had shrunk after the war. We flew the T6 (Harvards), P51 Mustangs and the de Havilland Vampire – one of the world's first jet aircraft. It was this last airplane type which became very significant for me some time later when I went to ask for a job at Boeing in Seattle.

At that time there were very few pilots who had ever flown jets and they also wanted ones with a degree in engineering. Although I could say yes to the jet aircraft experience, I didn't stress the point that it was one of the smallest jets in the world. It was powered by a jet engine and that's what they wanted."

Q. So it was straight onto Boeings?

"Yes, almost from the day of the interview, I was employed by Boeing. At first in Kansas working on the B-47 and then later back up at Seattle."

Q. You must have known Tex Johnston?

"I was there at the same time as Tex Johnston. He was Chief Test Pilot and was my boss. Of course, Tex was famous for flying the B707 through a barrel-roll over Lake Washington on Gold Cup Day during Sea Plane race week and he did it TWICE! It was quite low, about 1500 feet – he did it a second time because he thought people would not believe what they had seen him do the first-time round! He was a great stick and rudder pilot. When he was called into Bill Allen's office the morning after and Mr. Allen (President of Boeing Airplanes) asked him what he thought he was doing, his reply was typical, 'I was just selling airplanes Mr. Allen'. Either way Tex was instructed not to do it again.

Really it was a big risk because the 707 was our most important project at that time. The company had bet the bank on that one. It was similar years later to when the 747 became so very important to the manufacturer. This was just after Boeing had lost the contract bid to build the C5 for the military in '64 – it went to Lockheed. Up to that time nearly all our energies had been devoted to the C5 and the 747 was going along in the background. It was the airline Pan Am which kept the project supported because they were very keen to get the airplane.

They could see the commercial value of it and were continuously encouraging Boeing to produce it."

Q. So what happened at Boeing when the C5 bid failed?

"...well as soon as we didn't get the C5, immediately everything went into high gear on the 747. On 9th February 1969 - nearly 50 years ago - we flew the first B747. I was the company's Chief Test Pilot at that time, but I flew as First Officer for Jack Waddell – he had been slated for the first test flight from a long way back." (That airframe, 'The City of Everett' is now in the Museum of Flight at Boeing Field, Seattle).

Q. You had already done the test flying on the B737 by then?

"Yes... in April 1967 I did the first test flight of the '37. That one was not without issues."

Q. Did you know that the B737 was going to be such a winner? (They delivered the 10,000th airframe in March 2018).

"We really had no idea from the start that the airplane was going to be such a success. I recall we even felt the company was ready to give up on it at one stage. At the beginning it seemed like everything was wrong with that airplane. The Vee Speeds were way off and the stall speeds, way too high. The original reversers, didn't reverse, we had the main-gear shimmy (oscillating vibration) – oh we had one thing after another. And we corrected them all, one by one..."

Q. Was there any individual test flight which you recall...?

"Once we did have a big failure. I was doing the last point of the flutter testing. This is where you get the maximum speed, at the combination of both maximum Mach number at the same time you hit the maximum airspeed at about 16,000 feet. Well you can imagine we are way out beyond the red line there, because you had to give it a margin (for the flight envelope for normal operations). Well, all of a sudden, all Hell broke loose! The airplane pitched and then it yawed. It yawed so badly that one of the engines surged and I thought, engines don't surge at high speed..."

Brien Wygle and Lew Wallick 1967 – Boeing 737 testing.

Q. What went through your mind at a time like that?

"It rolled, pitched and I felt as if I had no control. I thought as it was a flutter test, we must have set up a flutter condition in the fin or the stabilizer and THAT part of the airplane has DISAPPEARED! Separated. What it was, we found out was the leading edge (extra lift) devices had come out, one here, another one there. Some came all the way out, some stayed back, some got stuck cockeyed. I throttled back right away and got control of the airplane again, but when we got it on the ground and we got out and looked at it... (all the leading-edge devices on the wings partially destroyed) These things were all hanging down – the airplane looked badly damaged."

Captain Brien Wygle interview with James McBride - Seattle 2018.

Q. So, not an easy birth with the 737?

"There were some real problems at the beginning and bit by bit we took care of them all. Then we were flying the -200 and that was a pretty good airplane. We had a very busy flight test schedule being run out of Boeing Field and the FAA Type certification was issued in December of that year."

Q. In the thirty plus years at Boeing, you were actually making aviation history as part of the team there, what were your key roles?

"I went through the jobs of Chief Test Pilot, Director of Flight Test and then finally Vice President of Flight Operations for Boeing."

Q. An impressive resumé for a farm kid from Alberta. What was your favourite job?

"Which job did I enjoy the most...? Without a doubt, Chief Test Pilot!"

Captain Brien Wygle died aged 96 in September 2020. He was an active supporter of the Boeing Museum of Flight in Seattle. His contribution over thirty years of testing, evaluating and selling airplanes should not be understated.

He had been a production co-pilot on the B-47 Stratojet, then lead project pilot on the B-52. Moving over to commercial in 1957, he did flight testing on the 707 then flew the very first B737 test flight. In 1966-67. Brien flew the B737 test programme all the way to type certification, and then flew on the crew of the first B747 flight. He retired in 1990, enjoying 30 years of retirement.

Captain Wygle was an active donor and supporter of Boeing's Museum of Flight.

10 IN A FLAP

By Capt Nick Francis

Some emergencies creep up on you. They unfold with a chain of events that can be recognised and hopefully countered by all those involved in operating and supporting a flight. Indeed the whole flight safety industry has grown and flourished based on a need for all of us to be vigilant for any link in a safety chain that could be faulty. This emergency didn't happen that way. This emergency started with one, enormous, 'Bang'!

I had just lowered the final stage of flaps on my Bandeirante and was working towards stabilising the threshold speed whilst feeling contented that it was a clear starlit night with light winds. An uneventful landing seemed assured. I was delivering yet more mail bags, single crew in the middle of the night whilst supporting Britain's Post Office. It was the 4th of August 1979, my wife's birthday. But back to the Bang.

The immediate result of this was to see the Luton approach lights rotate themselves around the windscreen of my 'Bandit' as it achieved a roll rate similar to a Pitts Special. As the angle of bank went through vertical I noticed the altimeter just going through 700' and the vertical speed indicator heading downwards, very rapidly! It is well known that the reaction of mankind to his imminent demise is 'Fight or Flee'.

As a flying instructor I faced many moments when students had tested this theory on me and in many imaginative ways. I am pleased to say that I have always subscribed to 'Fight' being the better option to realising my pension, thus it was to be this night. The roll to the left had now gone well over the vertical and was probably getting to around 110 degrees. Of course I had immediately applied full opposing aileron, but whilst it had slowed the initial roll rate down it was very evidently not going to be enough to prevent full inverted flight. Full right rudder followed and this helped a little more but now I was side slipping so badly that the increased drag of sideways flight only served to make the lights on the ground approach even more menacingly.

I figured that if these controls were to work at their best then airspeed would be vital. So along with full opposite aileron and rudder came full power on both engines. I smashed the thrust levers fully forward thinking that it was going to be a bit academic to worry about over-powering the

PT6's at this stage. Thus initially I thought I had done all that I could. The roll abated slowly, and for a few seconds I stabilised in a semi-inverted attitude. However whilst I had probably bought a few more seconds of thinking time, indeed a few more seconds of life. It was evident that I was going to have to do more to survive.

An Embraer EMB 110 Bandeirante – affectionately known as a "Bandit."

Photo Iain Mackenzie

Now as all commercial pilots and twin owners will know. At least once or twice a year a nice man sits alongside you and really spoils your day by pulling the power back on one engine just after you have made a nice takeoff. This produces a very violent reaction of roll and yaw towards the dead engine. Applying full opposite rudder and sufficient airspeed to make sure it stays effective, normally guarantees continued flight. So I figured if it takes a dead engine to spoil my day, then a dead engine right now might make my night a lot better. I snapped the right thrust lever closed and prayed the left engine would continue screaming its objections to me.

It worked. Slowly the roll reversed and then gathered momentum back to level flight. The 'Bandit' started to climb and I called 'going around'. It also seemed prudent to declare a 'Mayday' at this point and explain the circumstances to ATC. Once level in the circuit and downwind I started to take stock of the situation. I now faced the problem that to maintain level flight I had to keep the left engine at high power (by now within limits) allied to a large amount of right aileron with the right engine throttled back. In all I had a totally out of balance aircraft at a stabilised speed, but with very uncertain handling characteristics.

Whilst roll to the left was no problem, roll to the right was very limited indeed. It was somehow intuitive to do a left-hand circuit so at least no right turns were required. I could turn left and fly level. But the problem of landing under control was not going to be easy. It had been immediately evident that I had a flap failure on my hands.

The complications to this however centred on two areas. First was the Bandeirante's flap mechanism. Like many aircraft it had a central power source (in this case electrical) activating cables and rods to operate each flap. So whilst I had assumed that this was the source of the bang and I had taken action to reach over and select 'flaps up'; it had no effect. This was due to the fact that most sophisticated aircraft have a 'mutual lockout' designed to protect you against deploying asymmetric flap. The position of one flap is constantly monitored against the other and if a difference is detected then power is shut off preventing the situation from getting worse. Whilst this is fine when one flap fails to deploy or stops halfway down, it does not help if, like me, you had two flaps that had lowered perfectly, only for one to fail and retract without warning.

Having acquired one bad flap, the system had worked as designed and I was going to be stuck with my problem. The next few minutes were spent in experimentation. If the balance that had got me back to this stage was upset then I could expect dramatic results. Thus I joined that select band of ultimate aviators who explore the boundaries of an aircraft's operating envelope.

I had unwittingly become an experimental test pilot! I tried different engine settings, slowly reducing power on the left and force on the ailerons, then increasing the power on the right to see if more control could be achieved. As I tried different settings and powers it felt like the 'Bandit' was working with me to 'heal' itself. I felt as if the aircraft and I were truly working together to avoid what we both knew would be the result of failure. I spent most of the time in level flight way out of balance with the aircraft skidding to the right. I never did find a truly balanced condition, but I did find a compromise that got me round the circuit.

Gradually I found that I could maintain a reasonable degree of control with the speed a little above normal. I made a wide circuit and settled onto the final approach for a second time now seeing a multitude of blue flashing lights ahead of me. The emergency services had truly been awakened. The ground approached and it was time to compromise between reducing power on the left engine and accepting the inevitable left roll that would follow. Luckily with the right flap fully extended and gear down the drag was so high that the left engine would be required almost to the ground. So at the flare power was brought slowly back and 'God took control'. Starting the flare with a little right bank to allow for level wings at touchdown the landing was one of the smoothest I've ever made.

With the fire engines in hot pursuit I cleared to the apron and shut down. When I examined the aircraft the cause of my problem was only too evident. The right flap was stuck in the down position where the lock out system had sealed it. The left flap was also down, but this time loose on its hinges, free to travel down or up. The actuator had failed, shearing off inside its housing. Having extended, the failure allowed it to retract to the fully up position forced by the airstream. The mailbags were unloaded, and the company contacted to inform them that I would not be returning that night.

There was no engineering available at that late hour so all that remained was to secure the aircraft and retire to a hotel. Sadly the only room available was in a small one with a closed bar! Engineers arrived the next day and replaced the screwjack on the left flap. With the airframe and engines thoroughly checked I ferried the aircraft back to our base.

It transpired over the coming weeks that the left screwjack had been devoid of all internal lubrication. Normally packed with grease by the manufacturer, this one was totally dry. This was one of the first Bandeirantes operating in the UK and had more time than many on our fleet. So since being on the aircraft the screwjack had been corroding away until it decided to choose me as its victim.

As it was sealed I don't think that there was any procedure to open it up for inspection so of course no one would have had any inkling of the problem to come. I never saw it, but Embraer were reported to have sent a fax to the company offering their congratulations to me for being the only man alive to have flown a Bandeirante with one flap fully down and the other fully up. No one had reason to fly the aircraft in that configuration, but it was an honour that I could have lived without.

In my memory this flight highlighted the importance of knowing how to react to the unexpected and to appreciate the aerodynamics that keep us in the air. Recovery from unusual attitudes is now returning to airline training syllabus worldwide as events and accidents that could have been avoided have made the industry wake up to the problem of pilots losing control. Finally though, my overwhelming memory of this flight was a lesson in psychology. We all get tested to our personal limits. This night was my turn. I returned with the knowledge that whatever flying threw at me I knew one thing for sure. I would always keep fighting for survival all the way to the ground. As the saying goes,
"It ain't over 'till it's over."

But enough of the lecturing, let's return to that night. You may remember it was my wife's birthday. Not much relevant about that you may say. However as I set my bag down on the bed of my hotel room I noticed a card set by the phone obviously placed to encourage use of the hotel facilities.

I picked it up and read it with a wry smile. What did it say? 'Have you phoned to say you've arrived safely?' I rang my wife, and after wishing her a happy birthday, said,

"You'll never guess what happened to me tonight..."

I still have that card today.

11 SENTIMENTAL AIRLINE PILOTS?

By Capt James McBride

Commercial airline people are not known for their overt sentimentality. In fact they work in such a highly regulated industry that normally cool, factual, professionalism takes precedence over emotional hyperbole. It is refreshing and sometimes surprising then when evidence of our colleagues' gentler side makes an appearance – I believe this is referred to as being "fluffy."

In fact the very way in which the word "fluffy" is used by crew members indicates that it is an alien concept to most airline personnel. The reasons are manyfold, although chief among them must be that the industry is a very hard-edged business environment in which to operate.

This is an industry which thrives on TLAs (Three Letter Abbreviations) for just about everything and which has a language all of its own that is littered with jargon and technical expressions. From manufacturing, through commercial, all the way to flight operations, there is a continual reliance upon numbers. Everything has either a part number, a registration number, a stock reference number, a regulatory approval number and even the people are branded with employment numbers. Passenger e-tickets are given booking confirmation numbers, while crew tickets for deadheading (such a lovely expression) are given locator numbers, which have a similar purpose.

In Flight Operations there are numbers for speed, altitude, direction, weight (mass), while in Engineering it is a numerical feast with digits for every occasion available. In fact it is possible to have a face-to-face meeting between two airline employees where the spoken words are composed of virtually all numbers or esoteric expressions unfathomable to the outsider!

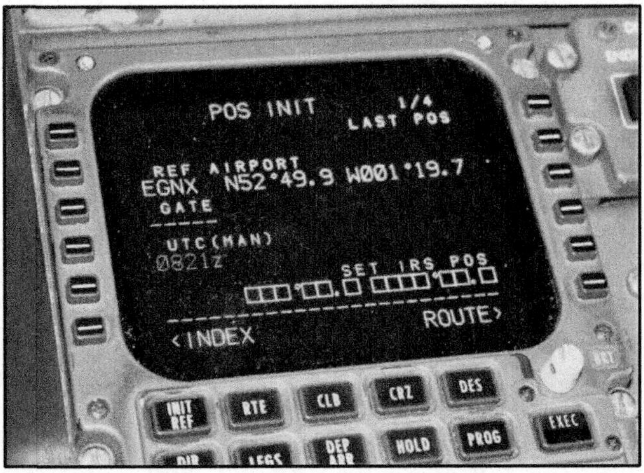

A High-tech industry... continual reliance upon numbers and TLAs.

The abbreviation DOB does not stand for Date of Birth anymore, but Death on Board; UnMin is short for Unaccompanied Minor; GPU is Ground Power Unit; ASU is Air Start Unit; LIAC is Late Inbound Aircraft when it comes to recording reasons for delay (or Code 93); PAX is short for Passengers; PAP is just one Passenger; CDL is Cabin Defects Log not be confused with CDU which is Computer Display Unit; FMC and FMS are very similar, but not to be confused and in the same vein, IRS's and IRU's are also very close in meaning...

It is possible that an FTL exceedance or FDM event requires an ASR to be filed which might become an MOR under the company's SMS when the FSO gets involved if they think it should be sent to the CAA.

Not all ASRs are managed in this way and that is why there is an SMS there in the first place. (Incidentally don't confuse an ASR with a PSR or CSR they are totally different things). Talking of FTLs, of course brings us to CAP371 and the strictures that it places on crew working patterns in that a crew member could be on SBY, CTBL when called out.

It might not be long before they reached their max FDP and went into Discretion which would have subsequent consequences for their Minimum Rest period to follow... Then the engineers are great when they start talking about ADD's and CRS's when it comes to filling in the Techlog.

Items such as Tyres become WTL, while engines are BSI'd or have MPAs while the MEL and DDG cover dispatch with unserviceable items which are then repaired at 'A' Checks, 'C' Checks or 'D' Checks. These repairs or replacements are recorded as being carried out IAW, AMM code XYZ-1234 etc.

Oh, and when it comes to old father time we really go to town! UTC is fairly straightforward, but then comes ETA's, STD's, STA's, ATA's and ATD's, then we have CTOT's, EAT's, EET's and so it goes on.

In the midst of all this tangled web of communication are the employees, *the people* who make it all happen. Somehow we manage to get our messages across to each other and at times the whole industry seems to run like clockwork – *amazing*. In fact there have been countless times when we (as a crew) have been rushing to try and get an aircraft away from the stand on time when it really has looked hopeless even up to just a few minutes before STD (Scheduled Time of Departure).

The holds are still open, the cabin crew are still counting heads as the last passengers are boarding while in the rear galley, there is organised chaos as the last catering is being done. The pushback tractor (which has been called for on the radio several times) is nowhere to be seen and the handling agent is making calls in the terminal for the missing passengers who have not yet found the gate...

All of a sudden, the pushback team appear round the end of the pier, as the catering truck pulls away and the rear galley is secured, the hold doors are closed, the last pax appear at the door and are ushered to their seats, the senior stewardess appears in the flightdeck door with a thumbs-up asking "Okay to close-up?" normally while you are halfway through a PA to the cabin and the First Officer is calling for Push and Start on the Ground frequency.

Then the checklists are all completed, and the groundcrew on the headset are saying "Release the parking brake please Sir, commencing pushback, you are clear to start engines 2 then 1..." You look at the clock and it is *exactly on the minute of STD* and you wonder, 'how the hell does that happen?'

Incredible, but true, we have seen it many times. There are so many factors which can go wrong to prevent the flight departing on time, but so often we manage to get away just on the minute – it really is very strange. Yet that also is part of the magic of the industry, in that although there is such a reliance upon technology you can always see and relate to the human factors in the mix. The politeness and good manners of people working under incredible pressure, the private jokes and funny expressions from crews which have bonded together – often they develop a sub-cultural life of their own.

I recall a crew years ago whom I met while downroute and they had all been together for over a week. We worked for the same airline and I knew some of them individually, but I had not been part of their 'gang' on that trip. They had found something incredibly amusing about the expression from the London underground system "Mind the Gap!" and this had become their in-joke. Whenever one of them said it out loud as if announcing it on a tube station platform, all of the rest of the crew would collapse laughing hysterically, this while others of us shook our heads in wonderment! But they were funny to watch.

There is also humour which crosses the divide between different sections of the airline community. Sometimes as flightdeck crew we witness this when talking to the dispatchers or handling agents, maybe even on the headset while communicating with the pushback crewchief. I remember once when we were downroute and we had switched on the red anti-collision lights before engine start on the ramp – there is one mounted on top and one underneath the fuselage. We could see the reflection of the one above, flashing red in the terminal windows, but I suspected that maybe the lower one was not working so I asked the crewchief,

"Is the red anti-collision light on down below?" the answer came back.

"It's On..... it's Off.....it's On.....it's Off....it's.." I interrupted him,

"Oh! Yeah thanks a lot Ground, very funny, we're cleared for push and start now!"

But we were still chuckling as we lined up for takeoff. Talking of which reminds me of Captain Dave Williams' experience while he was still flying

as a skipper for British Airways on the B767 fleet as they lined up for takeoff one day with *"landing traffic at four miles"* from ATC. His First Officer questioned him in French out of the blue, "...Sur le Croissant?", to which Dave replied "Eh, what?" and the First Officer explained,
"Are you going to take it on the *Roll*...?"

Yes funny times and funny people combined with long duty days and major time zone changes produce an emotional roller coaster of a working life. The great leveller in it all is the flying. No matter what your position, you are a key part of the team which makes it all happen and hence the current focus upon CRM (Crew Resource Management) to get the teams functioning together at a high level as they should.

While flying as a new First Officer on the B757 with my first airline, I had the nervous pleasure of flying with one of the most senior pilots in the company as my Captain. On the return sector an incident occurred which emphasised how well the senior pilot had managed the CRM to make me, as a very junior F/O, feel at home and able to speak my mind.

We were in the cruise and flying serenely above the Alps in the darkness heading back to Manchester when he said to me,

"So what's the latest gossip then? You've got your finger on the pulse James, what rumour's doing the rounds on Galley FM?

"Erm.... Nothing really Neil, it's all pretty quiet out there on the shopfloor", was my reply, but he was insistent and was not to be dissuaded easily.

"Go on, you can tell me, I like to hear what scuttlebutt is going around, it helps me to understand if the crews are happy or not, I like to know about the crew morale."

I could see that my stonewalling was not going to be enough and it was true that he was a good boss who liked to stay attuned to the mood of his people. There was one rumour though which had been circulating for some time now, maybe I should tell him...

"Well.... If you must know..."
"*Yes*?" he replied encouragingly, then he took a sip of tea, as I took a deep breath...
"The latest gossip is that... *you're having an affair with Emma Cousins!"*

I blurted it out, to which the reply from the left-hand seat was spluttering and choking as his tea went everywhere!
"Well you did ask...?!" I said, as he composed himself dabbing at his shirt and tie with a tissue.

"Nooooo... I know where *that* one's come from, erm we have a very similar sense of humour... and we were on a trip together to Florida last year... and...", I stopped him.

"It's okay Neil, you don't have to explain it to me you know, I am sure it's just an unfounded rumour", I said with a huge smile on my face, secure in the knowledge he would not be asking anymore questions about crewroom gossip on that flight.

So there is humour and humanity throughout the industry too, although it is rare to see it when we are all hard at work trying to be compliant with the regulations, while cutting costs and maximising profit margins.

I did see real evidence of our "fluffy" side though while viewing two Boeing 767s which were waiting to be purchased by a new owner. They had been delivered from Gulf Air in the middle east into the care of an engineering organisation while the sale of the airframes was completed and from an original fleet of 18, these were the last two which had left the airline. The very last one in fact was Golf Victor, being the last two registration letters of the airframe. The aircraft has been nicknamed "Vicky."

The engineer who was guiding us round the airliner pointed out there was graffiti on the hull in permanent marker pen to mark its last flight from the Gulf State where it had been flying for 15 years. The last crew were saying their goodbyes to an old friend and it was touching to see. By the Forward left entrance door, on the outside of the hull, there was *"goodbye my love goodbye"* and on the lower fuselage were written lots of warm messages such as *"Goodbye Vicky"* and *"Goodbye Sexy"*, *"Thanks for the service"*, *"Good luck!"* etc.

Even the local Bahrain CAA representative had signed a goodbye message and there were some in Arabic too. The aircraft had been in service with the airline since new, 15 years before and now it was time for a change of equipment, yet it appeared that the machine had developed something of a personality. Very soon of course those sentimental messages will be painted over by the new owner and the aeroplane will start a new life flying between new destinations in a brand new colour scheme.

After all it's only a machine, an inanimate, lifeless object without any capability for feelings or emotions, but just for a little while, it showed the tangible proof of the human (fluffy) side of our industry. It was a good feeling to have witnessed it.

It's only a machine an inanimate lifeless object without feelings...

Note: Emma Cousins was not her real name and it was only ever a rumour anyway...

12 YOU HAVE MAIL!

By Capt Nick Francis

"My God! It's raining lion's and Saint Bernard's out there!"

I looked up to see John, a fellow company pilot fully dressed in hat and overcoat looking as if he had just lost a fight with a fire hose shaking himself off as he closed the crewroom door. I steadied the coffee cup on the arm of my chair against an onslaught of the cold wind and rain as it attempted pursuit of my soaking colleague.

It was midnight in Liverpool 1980, and Britain's post office had woken up to the fact that mail could travel by regional airliner as reliably as it could by rail. Thus my company was one of many small airlines who found that money could be made by carrying mailbags by night and passengers by day. My trusty workhorse of that night was a DHC Twin Otter, or 'Twotter', the nearest thing to an airborne truck I've ever flown. As the laws of the time allowed us – it was flown single-crew without an autopilot. As night mail had by this time grown to be a large and busy industry competition among companies had grown fierce to perform on time and on cost in any weather. Thus the stage was set for me to succumb to that foul night and make one of the biggest mistakes of my career, one that I truly learnt from.

The post office had elected Liverpool to be a nightly hub for aircraft arriving from all over the UK. Mail was sorted, re-bagged and despatched onto its destination. Thus I had flown in from Glasgow, and having sprinted through the heavy rain into the pilots' crewroom had left the unfortunate baggage loaders to offload my 'Twotter' under the supervision of Patrick, our resident engineer. Now 'Pat' as we'll call him was a godsend on this run. He would make sure that the aircraft was offloaded and loaded in a manner which would comply with the company's instructions leaving the pilot to go in search of coffee and an hour's rest somewhere warmer and more comfortable. We all had good faith in him and trusted the system to work well, which indeed it usually did.

As my soaking colleague and I exchanged gossip on the latest company rumours Pat popped his head round the door and told me that the loading of the 'Twotter' was nearly finished and that time was going to be tight to get away on schedule. So it was on with the hat and Snorkel Parka (when you were based in Scotland you needed a 'real' overcoat) and it was my turn to do battle with the firehose.

Now I have never believed that just because it's raining heavily you should scrimp on your external walk round. Indeed I didn't believe it then. But that night whilst I scuttled around the 'Twotter' poking my torch beam into all the likely troublesome areas, cursing that I was going to spend the flight back to Glasgow extremely wet, I omitted to do one thing that I normally would have done and that had a lot to do with the subject of where things were.

The 'loading system' was to fill the cabin from floor to ceiling and front to back with mail bags in such a way that tying the bags down was unnecessary as they 'bulked out' the available space and therefore couldn't move. Then any lockers were filled evenly to balance up the aircraft, and the weight calculated to ensure a takeoff below maximum. This was where Pat was worth his weight in gold. It was his job to ensure that the Liverpool airport loaders followed this pattern, and we trusted the system to work. Big mistake!

So, suitably satisfied that the internal cabin lights could not be seen due to mail bags from floor to ceiling I climbed in, disgorged the soaking Parka and with Pat pointing at his watch asked for start clearance. The tower approved me to taxi, so it was brakes off and away for the end of Runway 27. In those days Liverpool's present terminal had not been built, but its new runway had, so we had to taxy all the way from the old terminal some two miles or so to the end of the runway in use. It was at this point that I got the first message from the good lord that all was not well in my world. As we all know; when the good lord speaks, one should listen.

As I taxied along I became aware that the aircraft nose wheel was very light on the ground. Steering, which was hydraulic via a large 'phallic symbol' jutting out of the control column seemed ok. But every now and again I became aware that the nose wheel was riding very lightly on its tyre. I should of course have started to think why, but given the time schedule and the thought of returning in the inclement weather, I was inclined to put the bouncing of the wheel down to the wind that by now was blowing behind me. Anyway I convinced myself that as the nose wheel seemed fine when I selected reverse thrust then not much could be wrong.

Onto the runway and it was wipers on, full power, and off into the driving rain and low cloud. The thought that I was on my way home to a waiting bed made the squelching of my socks seem less important. Indeed the flight up to Glasgow was uneventful as far as the climb and cruise were concerned. The memory of the odd nose wheel was replaced by the comforting fact that the aircraft was handling normally. The cold front that was drenching Liverpool had long since cleared Glasgow, so I was cheered by the Volmet telling me that Glasgow was by now enjoying a partly cloudy sky with the traditional 'no excuse' wind of ten knots down the runway.

Now when you earn your living flying a turboprop around some of Scotland's and Scandinavia's finest selection of foul weather, you get to be a little short on 'fun' activities that boost your morale. Don't get me wrong, I enjoyed my job, especially the day schedules. But five consecutive nights on the mail run left you a bit short of gaiety! Consequently one of our 'games' used to be to chase Britannia Airway's Boeing 737's around the radar vectoring circuit to the point where the controller had to tell us to slow down.

Okay, I know it's pathetic, but when the 'Twotter' was flat out down a cliff with the wind behind, it would just about match a 737 with intermediate flap. Thus as luck would have it that night I slotted in number two to an incoming 737 and wound the old 'Twotter's' elastic bands up as far as they would go!

It was at this point that fate delivered the necessary input to the safety chain that was to make this night more than eventful.

As I passed six thousand feet I encountered a couple of small cumulus clouds. Nothing very large; probably only a couple of thousand feet tall. The visibility was good, and they seemed to be a lurking pair of solitary vagabonds wandering the sky on their own. I clipped the left of the pair and felt an inappropriate amount of turbulence bounce me around. As I retrimmed I felt and heard a resounding 'thump' emanate from the rear of the aircraft. My initial thoughts centred on the fact that I must have hit the 737's wake, or the upper winds must have been more northerly and I had hit some turbulence downstream of nearby hills. The 'thump' quickly resolved itself and the 'Twotter' was once again flying smoothly. Glasgow approach asked me for my speed and childish satisfaction was experienced when they asked me to reduce to avoid closing on the 737 that was now on final approach. Swinging in behind him for a visual approach I started to concentrate on reducing speed for the landing and it was at this point that I started to reap the rewards of my mistakes.

As I reduced speed for the approach heading towards a hundred knots I became aware of the fact that I was trimming forward an awful lot, indeed at around one twenty I felt the trimmer handle reach its forward stop. *Funny*, I thought, *I must have got a fouled cable.* No problem, I would just have to land out of trim. But oh no! Things really became obvious when at around a hundred knots I ran out of forward elevator! The control column reached its stop and I watched with horror as the nose continued to rise and the speed fall rapidly towards areas of the Air Speed Indicator that I did not wish to be acquainted with. Suddenly all the pieces of the safety chain fell into place. The centre of gravity must have been well aft to cause the nose wheel to be skipping lightly whilst taxying at Liverpool. The 'thump' I had felt passing the cumulus must have been due to the mailbags shifting in the main cabin, and no, I had not got a jammed trimmer. I had a Twin Otter on short final at Glasgow with its centre of gravity located somewhere near Edinburgh!

Obviously a go around seemed to be the first priority, but that was only going to delay the inevitable. However it certainly seemed the immediate option. After all, if I had control up to this point then retreat seemed the best course of action. I then wondered if retracting the flaps would be safe, and indeed should I try a flapless landing next time. These thoughts were quickly dumped without answer as the falling speed was by now going to regions known only to helicopters! Luckily as I gently increased power to start the go around the nose fell back towards the runway and I realised that with the 'Twotters' high thrust line, and the tail plane being almost in line with the engines, the increased airflow from the props would at least give me more control over the elevators, sufficient at least to attain level flight by the runway. The best option then seemed to be to fly the aircraft onto the runway fast and flat. Then with forward elevator combined with brakes and reverse thrust giving a nose down pitch I could expect to keep nose wheel control whilst I reduced speed to a level that would not cause damage to the aircraft once it was left to its own devices.

So with enough thrust set to give me elevator control I flew the 'Twotter' onto the ground. I selected reverse, pushed forward and using brakes decelerated quickly to a walking pace. As I twisted the throttle grips back to idle power the 'Twotter' promptly sat gently on its tailskid and proceeded to rock slowly back and forward between the skid and its nose wheel. A perfect seesaw.

Re-applying power seemed to bring the nose down and I was able to get enough steering to clear the runway and taxy to the ramp. Luckily in a way this was happening at night and air traffic were not party to the site of a Twin Otter imitating a nodding dog on their runway. The whole event lasted less than two minutes. There was no damage so I felt the best thing to do was to clear off to the general aviation park and seek out a very strong coffee!

As I taxied into the apron, a handling agent stood waiting for me to draw up. Somehow I had a premonition that he was going to open the main cabin door, so after shutting down I pulled down the side window and yelled,

'"Whatever you do, don't open the door." He later told me that that he didn't get the first four words of that. So I was alarmed but nevertheless unfairly amused at the sight of a pair of legs sticking out from under a pile of around twenty mail bags as they flattened him to the apron.

It was then that all was revealed. You will remember that the 'system' was to load the cabin front to back, floor to ceiling, then fill the lockers. Indeed the cabin was full floor to ceiling, but only for two thirds of the cabin, leaving around six feet of free space for the bags to move during the turbulence. And the lockers? Oh the back one was full to the brim, but the front one was completely empty.

'The System' had failed partly due to the fact that Patrick had not unreasonably taken a comfort break during the loading. The loaders were more used to seeing a Bandeirante on this run and that didn't have a nose locker. But of course the main culprit was a pilot more interested in getting airborne than double-checking what he trusted would be done on his behalf.

Twenty years later and I now have to walk a little further around the perimeter of a 767. Not one single walk round occurs where I don't remember this incident and I always strain my neck to look as far as possible into the cargo areas for any odd additions to the load. Doing this over the years has revealed many a surprise varying from stray main wheels to pallets of Mangos.

My mind always recalls that night, and how lucky I am to be alive after it. Whilst I may write about it jovially now, at the time it was a salutary lesson for a young man building his career on how easy it is to come to grief in Aviation.

A Twin Otter - similar to the one in "You have Mail!"

Photo Tom W Smith

They used to describe the Shorts 360 as "The box the Twin Otter came in," Ed.

13 BAD WEATHER? ...GOOD NEWS!

By Capt James McBride

I recall sitting in his office across from his desk, my new Fleet Manager for the B767 Cargo operation which I had just joined. He was explaining to me about the operation and how it all worked. This would be my first experience of freighting with heavy jet transport aircraft and I had lots of questions. It all sounded quite complicated and I shared my concerns with my new boss. He was anxious to assure me that all would be well as he looked at me steadily over the top of his half-moons and said with a straight face,

"Look. The important thing to remember about freighting James; is, we've never had a complaint letter from a parcel!" then roared laughing at his own joke. The interview continued in a similar vein and I left his office with his last words ringing in my ears, ."..and remember you won't need to do any PAs."

I could have been forgiven for thinking then that my new job was going to be a piece of cake, certainly that was how it came across from the boss. The only slightly disturbing aspect for me was the list of destinations to which we carried our freight. Many of these were in areas of either civil unrest or active war zones. You might be wondering what a commercial operator would be doing flying into and out of a war zone. Sadly, the fact is that War is viewed as Commerce by many politicians and their associates. This is the reality of the world in which we live.

Airlifting supplies for the military is a lucrative business and the freight airlines have never been busier. There are risks however and the danger of being shot down either accidentally or on purpose is real. Consider the DHL cargo Airbus A300 in Baghdad back in November 2003 which was brought down by a shoulder launched Surface to Air Missile by terrorists. The crew subsequently lost all hydraulic systems and literally landed on a wing and a prayer.

Flying these sort of missions is not for the faint-hearted. Certainly Baghdad was always a challenge for operators with a steep/tight spiral descent into the airfield which avoided leaving the airliner exposed to potential ground attack from outside the perimeter of the airport. My good friend Arthur, a B727 Captain once told me how he decided that flying the spiral descent arrival into Baghdad was not going to be part of his work pattern anymore. He said,

"So I realised I was actually risking my life carrying out these freighter missions into Baghdad for the US Military. It really hit me one day when I asked what the load was. It consisted of thirty-eight tonnes of Gatorade!" This was the last straw for Arthur as he went to see his Chief Pilot in Bahrain to resign.

You could see it from his point of view. All of that stress and angst for 38 tonnes of Gatorade...?? Imagine being shot down carrying THAT payload? Hardly a heroic end to your life, short and sweet though it may have been...

Silkway West Boeing 767 Freighter –"... you won't need to do any PAs."

Which brings me back to my particular situation, flying cargo Boeing 767s into and out of places like Kandahar, Kabul and Bagram. Names to conjure with for sure and none of them ending in –stan or –bad. An old journalist friend of mine once said that he .”..tried to avoid being sent to places ending in -stan or -bad.” He had literally travelled the world and over the years had developed a few simple rules to live by.

I arrived at the briefing room a few minutes after the First Officer, he had already printed off the weather reports and Notams* for the flight. We greeted each other with a formal handshake and then he smiled broadly and said, "Good News Captain! *The weather is BAD today.* Low cloud and poor visibility at destination! Our only problem is that I'm having difficulty printing out the return flightplan."

Of course as an airline pilot, usually bad weather is bad news, but the funny thing about this particular contract was that we knew our arrival at destination would be much safer for us in bad weather. We were flight planned to go to Kabul in Afghanistan with our Boeing 767 freighter carrying vital supplies for the military. Some of them were Dangerous Goods. The main risk we faced during the approach to the airport was that it was over high ground a long way from the field. Ideal terrain for the bad guys to take potshots at large targets like ours – as part of our standard procedures we always inspected the airframe after landing for holes...

In addition we had other issues to think about. Fuel was at a premium in Kabul and we tried not to refuel there – it was expensive and difficult to arrange. Whenever possible we carried roundtrip fuel to the airport so we could fly in and out with minimum ground time. This would mean we would be landing at Maximum Landing Weight – and then we tried not to use the brakes after landing.

The reason for this was to reduce the brake cooling time required on the ground. We used maximum reverse thrust to slow us down on the very long runway prior to vacating at the end. *Not for the faint hearted* often went through my mind as we commenced our descent into mountainous terrain, heavily laden, planning for a minimum ground time turnround.

On this day, we broke cloud at about 500 feet above the runway – very pleasing. The landing went well and we rolled right to the end of the 11,473 feet runway turning off just past the wreckage of the Boeing 737 freighter which had wiped out the ILS system only a week earlier. They had misjudged their landing and ran off the end with unfortunate consequences. The airport had only just completed the long-awaited installation of the ILS a fortnight before. Now the Notams read "RW29

Parked on the MIL Ramp - all sorts of flying machines taxy past us.

ILS out of service" again.

Our problems were not over though as we shut down on the ramp and the US Military commenced unloading the cargo pallets with their high-loaders and trucks.

On the ramp in Kabul "You're NOT leaving the aircraft are YOU!?"

We had arrived from Dubai without the return flightplan and now we needed to get it printed. I said, "No problem, I will pop over to the Ops building and get it done, you supervise the offload and I will be back in 15 minutes."

He appeared shocked and his voice raised an octave as he replied,

"You're NOT leaving the aircraft are YOU!?" I was surprised, but assured him I would only be a few minutes and back as soon as possible. I called the dispatcher and got a ride to the Operations building in the Follow-Me pickup truck. Once there I found another issue – how to get in? Everywhere there was a mass of armoured fences, CCTV and code pads on doors/gates. It took a few minutes to convince them to let me in and I had only driven FROM airside!

Once inside I quickly explained that I needed to call our Airline Operations Department and the United Nations peacekeeping soldier was very obliging. A short while later the flightplan came through on email. We printed it off. It was while it was printing that I noticed one of the nearby screens had come up with a large print message.

At the top of the screen it was labelled *TOP SECRET* warning of dire consequences for the casual observer, however the message did draw my eyes to it:

WARNING – SUSPECT IED/PKG FRONT STEPS BLDG 509. ALL PERSONNEL STEER CLEAR, MAINTAIN MINIMUM 200 METERS DISTANCE. BOMB SQUAD ON THE WAY.

I had got the flightplan and was ready to leave, I paused at the door,

"Tell me my friend, is building 509 far from here...?."

"No, it's next door, this is building 510... why?"

"Well, you might want to read the secret message on that screen there mate... *Ciao!"* I resisted the temptation to run. Of course it was more likely to be an innocent package than not, but what did I know?

Outside, the sirens started to wail with that eerie hi-lo wavering note, similar to air raid warning sirens the world over. Not a good sound, seriously not a good sound to hear at an airfield in a war zone.

Don't run... The bomb squad will get it done. Unless of course it is a device with remote control just waiting for them to turn up. I knew from reading about Iraq how the terrorists placed high value on blowing up the IED Disposal Team as they arrived on-site to do their work...

A few minutes later, back at the now largely empty freighter my young colleague was having kittens.

"CHRIST! WHAT'S THAT NOISE? IT'S AN AIR RAID!" I tried to calm him down having heard the sound change to a continuous whine.

"No don't worry Joe, that's the 'All Clear', the Bomb Squad have done their job – it's all okay now, but let's get out of here eh?"

There is nothing more exhilarating than the *'rush'* of a full-power takeoff in an empty (lightweight) Boeing 767 with a steep climb at a high angle of attack followed by a full-power climb into the blue sky above a solid overcast. Believe me. Seriously, it's the best feeling!

*

14 THE STREET FIGHTER

By Capt William Hagan

Captain William Hagan.

Much has been written about the attempted hijack of the B747-400 which occurred in the middle of the night on 28th December 2000. BA flight 2069 from London, (Gatwick) to Nairobi was cruising at 35,000 feet; most of the 379 passengers were asleep. In the cockpit alone was one pilot at the controls, Senior First Officer Phil Watson.

At 0500am one of the passengers, Paul Mukonyi, a well-built 27-year-old Nigerian entered the flightdeck quietly and closed the door behind him. Despite the protests of SFO Watson, "You're not allowed in here, you must leave, Get Out!" Mukonyi rushed towards him, overpowering the seated pilot and grabbing the control column. He leaned right across the centre pedestal, with his upper body across Watson's lap and proceeded to use force on the control wheel with apparent malevolent intent.

The pilot's view of the instruments was obscured, the Autopilot disengaged due to overriding force on the controls and the aircraft started a rapid climb as Mukonyi pulled back hard...

First Officer Watson knew that the airspeed was dropping dangerously low, despite the Auto-throttle system commanding the engines to full power. What he couldn't see was the pitch attitude, which was 20o nose up and the airliner was approaching 41,000 feet.

It was just about to stall for the first time. In addition to the pitch attitude, the airliner was rolled from side to side in the climb as Watson tried to wrestle back control. In the cabin the passengers were aware of the attitude changes, but more than that they were awakened by heavy airframe buffeting and shaking, which none of them had ever experienced before – people started to panic. As the aircraft stalled, the nose drops and the negative 'G' forces were felt in the cabin as loose articles, including passengers, started to fly around. In the lavatories, passengers were thrown against the ceiling and one of the cabin crew sustained a broken ankle when a bar trolley fell on top of her... Now grown men were screaming, people were praying and in the words of one passenger - *"This was the scene of an aeroplane about to crash"...*

However there is more to the event than the reports in the media. Captain William Hagan explains from the pilot's perspective. His story is extraordinary.

Q. There were some newspaper headlines which hailed you as a hero, how did you feel about that?

"Well, as you know, Captains get all the glory or all the blame. I would say First Officer Phil Watson was the hero really. I only behaved in the mode of a Street Fighter!" *(A reference to the fight in the cockpit between Capt Hagan and the intruder).*

Q. Was it fate that put you on the flight that day?

"No, it was the bidline (the pilots work bidding system) which put me on the flight that day. I nearly missed it. The weather in the UK was awful in the days leading up to the flight and my travel arrangements were disrupted. I was rushing to make it; I called crewing on the mobile to tell them my transport problems and asked if they wanted to put someone else onboard as Captain. They said, 'there is no-one else'."

Q. So it had been a normal flight up until then?
"Yes we had flown for 6 hours and I was due to rest for an hour or so before landing, which was why I was in the bunk. I deliberately remained in the bunk for some time, maybe tens of seconds, feeling the motion of the aircraft. I knew, or I thought I knew, that I had 2 perfectly capable, experienced pilots on the flightdeck. I thought, if I went in there and the aircraft is behaving erratically, I might make the situation worse. I could fall across the thrust levers or something – there was no seat for me to sit in to fly the aeroplane. *(What Capt Hagan was unaware of was that the third pilot had left the cockpit and was in the lavatory).*

Q. What made you decide to leave the bunk?
"Well, I took my earplugs out and I could hear Phil's voice shouting 'Help! Help! .....HELP!' – then I knew something was badly wrong. At that moment I could feel the buffeting and knew we were stalling with the engines at full power. I couldn't understand why we would be at 25 degrees nose-up and bank steeply from side to side. I even thought we might have hit space debris, but that didn't add up because I had not felt a collision..."

Q. When you entered the flightdeck, did you take long to understand what was happening?
"It was obvious to me straightaway that the threat to the aircraft was posed by an intruder, whom I assumed to be a terrorist. He looked like a terrorist and I even thought he had a balaclava on which he didn't. I thought about getting a weapon to attack him, but I thought it would take too many seconds time was of the essence! I went straight at him. I hit him a few times, but there was no reaction. It was like hitting a stone bollard, it had no effect. He was like Mike Tyson."

Q. What did you do then?
"I grabbed his clothing, but it was slippery and I couldn't hold on to anything. We were diving to the left at this point and the speed was increasing rapidly, I knew we were losing a lot of height. I grabbed him under the armpits and tried to lift him up; he was heavy and still hanging on to the yoke. Phil was shouting GET HIM OFF!"

Q. What were you thinking at this point?
"There was a lot going on in my mind. I didn't know what the altitude was now and I was trying to remember what part of the world we were flying over. I had just woken up and on many of my flights (to Australia) would have been in the bunk over the Himalayas and it would have been dark at that time also. But there was a little dawn light appearing in the sky and suddenly – I put it all together... *We're going to Africa!"* But then I remembered also that my wife and two children were onboard..."

Q. Did this affect your feelings at that time?
"Well, I thought we are going to Africa, THAT'S GREAT, the MSA* is not much more than 10,000 feet on our route. I've got more time. But when I thought, *I've got the family with me;* I had a mixture of emotions and that turned to anger that he was trying to kill my wife and kids. Also I felt suddenly ashamed, very ashamed that I'd let the passengers down. They were all going to die and I was responsible. All these years of flying, I've thought, well if you look after the cockpit, you're looking after the aeroplane. Now I was overcome with the responsibility and the feeling of shame for not carrying that responsibility through properly."

Q. Yes, I guess we don't normally think about it much as airline pilots, 'their lives in our hands' were you afraid?
"No I didn't feel fear – I just had this thought... *YOU'VE GOT TO DO THIS; YOU'VE GOT TO SUCCEED! AND... YOU'LL KEEP GOING UNTIL YOU SUCCEED...* and that's when I got the motivation to stick my finger in his eye. I did it really hard, pushing my finger in as much as I could. At that moment... if I could have killed him, I would have done if I thought it necessary. I felt so strongly."

Q. At this time others came to help you?
"Yes the 3rd pilot came in halfway through the encounter and then we shouted through the open cockpit door, "COME IN, SOMEBODY COME IN!" and 3 passengers came in – together we managed to get the intruder out of the cockpit into the cabin. He struggled all the way."

Q. And then the FO managed to recover the aircraft – that must have taken some skill to do...?

"Yes he did a great job, but then Phil would tell you that in his flying career he's spent much of it upside down. He was a military pilot before and did a lot of aerobatics, although never before in a Jumbo Jet – he is no stranger to unusual attitudes. We were at 37° pitch attitude below the horizon, with the speed accelerating very rapidly – he had to take care in the recovery. He flew it magnificently – at one point we were descending at 30,000 ft per minute, we lost 12,000 feet in 25 seconds. If we had gone supersonic the aircraft could have started to break up, he had to be extremely careful and precise with his control inputs."

Q. Presumably you have listened to the Cockpit Voice Recording of the incident?

"Technically no, there was nothing on the CVR! You would have heard Phil saying 'you're not allowed in here, you must leave' but the intruder said nothing at all. Then you would hear Phil shouting "HELP HELP, HELP!' which was the time I took my earplugs out. After which you would hear Phil shouting continually, 'GET HIM OFF! GET HIM OFF! GET HIM OFF!'"

Q. At what point did you address the passengers?

"Well, very soon afterwards. We immediately made a call on the VHF radio to advise other air traffic of our lost altitude, then I thought I must say something to the passengers and crew. It was my duty to reassure people they were not going to die. I said *'A bad man came in the cockpit and tried to kill us all, but everything is alright now'.*

Q. There is a short video on the internet, it shows you outside the flightdeck later, retrieving your shoe. By this time you had made your PA, so you must have made it with only one shoe?

"I actually made that PA in my underpants! Because the flight was full, my suitcase had gone in the hold, so I didn't have my sleepsuit with me. I didn't have time to retrieve it from my case."

Q. And on arrival in Nairobi, did you all have a survivors'

party?

"No – far from it. There were some injured passengers and a crew member who needed hospital attention and I also had to go to hospital. I had suffered a couple of wounds in the struggle on the flightdeck, my finger had been bitten badly and I had left some skin from my ear on the overhead panel. At the time we were all concerned about the possibility of AIDS. We landed at 7am, we were at the hotel by 10 and the police would not let us leave the country without interviewing us."

Q. So your visit to Nairobi was cut short then?

"Very much so – we were on the flight back home to London that night. The company suggested that it would be best for us all to come back straight away."

Q. What did you put in the aircraft Techlog after the flight?

"It was interesting; my first entry in the Techlog was to sign-off the De-Icing which had been carried out in London Gatwick prior to our departure. It was the procedure at that time for the crew to sign it off after the flight, because as soon as the de-icing was completed we would taxi for take-off. That entry in the Techlog was a reassurance for all of us, in the flightdeck, that we were back in the real world. After what we had been through it was a relief to write it. Then I wrote in the Techlog 'Please carry out Heavy Landing Check and any other checks necessary as aircraft was FLOWN OUTSIDE NORMAL PARAMETERS'. The engineers spent five hours checking the aircraft over and didn't find anything wrong with it. Fortunately most of the unusual attitudes occurred at low speed. The senior people in London were also very keen to know if the aircraft had exceeded any high-speed limits."

28-29 DEC 2000	BA2069	LGW	NBO	23.02	07.18	08.16 L	BNLM	11.12	HIJACKED
29-30 DEC	PAX	NBO	LGW	20.23	04.56	00.00	LM	PAX 8.33	ROBERTSON

A copy of the entry in Captain Hagan's personal flying logbook.

Q. In the debrief, were you all sure of the intruder's intentions?

"Yes, definitely. Phil said he was trying to turn the aircraft upside down and if he had done that we would not have recovered. Without a doubt he was trying to crash the aeroplane and kill everyone onboard. If he, Mukonyi, had been on the controls for a few more seconds and Phil had been unable to prevent further increase in bank angle I am sure he would have achieved that. An inverted stall would have been unrecoverable."

Q. How did it change your life and that of your family after the event?

"Firstly it made us all realise the fragility of life and the terrible thought that the lives of our young children could have been snuffed out so easily. Afterwards, at home it made us value the time we had together, to not fight about 'who left the milk out of the fridge' etc. In fact we didn't have any arguments in our household for a number of months if not years! It brought us closer together as a family."

"That Techlog entry was a reassurance... we were back in the real world."

15 ROGUE ELEPHANT

By Capt James McBride

I remember quite clearly where I was when I answered the call on my company mobile. It was a Saturday morning, the sun was shining and I was at a motor racing circuit enjoying my time off with some like- minded friends. From what I have seen of professional aviators over the years, we all need something entirely separate from 'the job' in order to de-stress, unwind, take time-out etc.

These activities can vary from hill-walking, horse-riding, through to crafting violins. In my case at that time it was a motorcycle track day. Of course I can imagine that some would say riding sports motorcycles at high speed on a racetrack is not very relaxing and there is some truth in that, however the important point is that it is completely different to commercial aviation. I looked at the phone number which appeared on my screen and recognised it as one of the airline's Line Training Captains at my base where I was the Base Captain - I was his line manager and also manager for all the other 100+ pilots. His voice was concerned and genuine, albeit hesitant. Nobody likes to be the bearer of bad news after all...

On the Monday morning, back in the office, I knew I had a serious problem to deal with, but wasn't sure where to start. The Line Training Captain had passed on a report (secondhand) from one of our First officers that "...there is a Captain flying the line in our base in an irresponsible and unsafe manner, he fails to follow SOPs, operates dangerously and will not listen to his Co-pilots." Naturally my first thought was to ask to see Captain X and confront him with the report which I had been given and ask him if it was true. Why not? The trouble was that if his performance was as bad as the verbal, anonymous report said then he would of course deny it and I as the Line Manager would be left with a decision to make – who to believe?

Also, in the interests of fairness and transparency, why should I act upon an anonymous, verbal report...? So I decided to find the

FO and speak to him confidentially. Just one call was all it took for me to find his name and I called him to come and see me next time he was in the office. Next morning he was due to fly, so he dropped in prior to his duty starting. He was an experienced FO whom I knew was a good aviator and was not far from being promoted.

He didn't want to "make any waves" in his own words, but he thought I should know that Captain X was operating well outside of the Standard Operating Procedures. Before he spoke, I offered him my thoughts, cautioning that all pilots were guilty at times of not operating *exactly in accordance with* the strict letter of the SOPs and that in some cases this may make us (as the other pilot onboard) feel uncomfortable. That alone was not, strictly speaking dangerous. I said that to accuse one of our colleagues of "unsafe" practices was a very serious matter and we had to be very sure of our ground before doing so.

Captain X flew the aircraft at maximum speed along the airways, being perilously close to the high-speed buffet, then descended at very high speed even in turbulent conditions.

He then proceeded to inform me that Captain X routinely flew the aircraft manually into 4 reds (very low on the PAPIs) on final approach to land as short as possible, using heavy manual braking to achieve the first rapid exit from the runway, sometimes taxiing onto the stand so fast that he shut down the engines before the FO had managed to retract the flaps...

Also, Captain X flew the aircraft at maximum speed along the airways, being perilously close to the high-speed buffet, then descended at very high speed even in turbulent conditions with no thought for passenger comfort or mechanical empathy for the airframe! There were some other serious deviations from the standard operating regime which he informed me about also. Perhaps the most worrying aspect was that when the FO said anything to remind the Captain of the SOPs, he was overruled allegedly in the most rude and arrogant fashion. The captain was a very experienced operator with many thousands of hours in command – so there existed a very steep authority gradient in his cockpit. Yes, this was very odd behaviour by a professional airline Commander to be sure – what to do?

Bear in mind, this was in the days before Operational Flight Data Monitoring (OFDM) which was just on the verge of being introduced in our company. Although the flight data was there in our fleet of modern Boeings, it was not routinely captured nor monitored. I had only the one report and it was verbal. It sounded bad, but again it was one word against another, I would have to investigate further. I asked if the FO would put it in writing, he was reluctant to do so even when I explained that I was unable to act on a word-of-mouth report. I thought for a moment and said that if I managed to find other first officers who reported similarly and were prepared to put it in writing would he do so then? He agreed.

Over the next couple of days, I spoke with 2 other pilots who had flown recently with X – they both confirmed the first report and worse. Both consented to put their name to written statements, but I still worried there was nothing tangible, a lack of real data. I took advice from our senior Flight Operations Manager at the airline's headquarters – he confirmed my own thoughts. I was not permitted to download the flight recorder data from any of Captain X's flights unless I had a report of a safety related incident on a specific flight which warranted such action.

I spent some time going through the company's operations manuals to be sure of my ground and also to ensure that we were being fair and consistent with Captain X in the application of the rules. I had actually done some of X's Line Training with him some months earlier and found him, at that time, to be a very competent, standard operator. I was surprised about the reports, but I also knew that the reporters were genuine in what they said. Independently each had given an almost identical narration of what constituted Gross Misconduct, worse still the miscreant was still on the flying roster flying aeroplanes in my base... I had to work faster.

My one big worry was that I had no firm data, no written or recorded 'evidence' to show – it was all 'reporting', whistle-blowers united, all saying the same thing, but no *documents*. I racked my brains for what else I could do. There was no way I could obtain flight data information without a real reason for requisitioning the recording. To do otherwise would break with established protocol and as a responsible manager I was not prepared to do that.

On Wednesday afternoon I was sitting in the office scratching my head, wondering where to look next when one of the FOs came by. We started chatting quietly about the case (an open office policy has its drawbacks) and I asked him again about some aspects of the subject Captain's operation of the aircraft. "How can I prove that he's flying the aeroplane so far away from the SOP? For example, his Cruise speed, you say he's flying faster, how much faster than ECON?"

The reply was quite specific, "He flies at Mach .80 in the cruise all the time; everywhere; burning loads more fuel and risking a high-speed stall..." I had an idea and said so,

"Well that will show up on the Engine Trend Monitoring Log surely?" I was pleased, now maybe I could see something tangible to support what all the co-pilots were saying. He smiled,

"...you won't find it there!" And then I realised. The entries in the Engine Trend Monitoring logbooks were actually written in by the aircraft Commander, inflight, with auto-throttle temporarily disengaged after a certain period of steady state flight. All of the engine parameters were then written down by hand, including the speed of the aircraft at that point and the cruising altitude, OAT etc. I guessed the reason why I wouldn't find the proof there and said so,

"Of course! He slows the aeroplane down to ECON cruise at Mach .75, records the readings on the engine instruments then goes back up to his previous speed again.... Right?" It was obvious to me, but he shook his head before replying.

"No it's even simpler than that, he doesn't slow down at all, he doesn't DO any engine readings, he simply copies the recordings from a previous page!" I was shocked. The Engine Trend Monitoring was an engineering/flight safety programme which was used by the engineering team to watch the health of our engines in the fleet. For an aircraft Commander to falsify the recordings in the book just made no sense at all apart from confirmation that he was totally irresponsible. The FO backed up what I was thinking by saying, "If you can prove that he's copied previous readings, you've got him!"

I had a Rogue Elephant and I had to stop him before there was a serious incident or worse...

That was Wednesday. My errant (law breaking) Captain was still flying the line and due to finish his week on the Friday, prior to a long weekend. I had a Rogue Elephant* and I had to stop him before there was a serious incident or worse. I went to bed that night concerned that I only had a limited amount of time to find more evidence before I would have to confront him anyway with what I had and stop him from flying irresponsibly. It would be a mess of course, I knew that. He would deny it absolutely and claim victimization, I could suspend him from the line for a small amount of time, a few days perhaps while I completed my investigation, he would claim "a conspiracy of junior co-pilots", "collusion with management", the unions would become involved... Dear God, it was a nightmare and I had to operate on the line myself the next morning...

I woke up well before my alarm went off and drove to the airport ridiculously early for my flights, it was still dark. On arrival instead of going to the crewroom to report for duty I had a mission. I would replace all of the Engine Trend Monitoring logs on all the aircraft. The night before while driving home, I had spoken to the Chief Engineer in the base and arranged for the on-duty early morning shift to have new logbooks ready for me.

There were 7 aircraft out on the apron in front of the terminal and I spent the next 20 minutes opening each one up, going in the cockpit replacing the logbook and closing them up again. The last one I didn't need to close as the crew were approaching from the terminal. The pilots raised their eyebrows to see 'the boss' out on the line so early, but I wished them a good flight with a cheery wave as I dashed past heading for my own report time at the crewroom.

There were 7 aircraft out on the apron in front of the terminal.

After my flights were finished I had time to sit in the office with the logbooks – the top copies had all been pulled from previous flights by the engineers, but the carbon copied pink slips were all present and legible. Each one had the name and signature of the aircraft Commander at the bottom - the books went back several weeks. I found Captain X's entries easily enough and all were in his neat handwriting, appearing very professional. For a moment I had real doubts about what I was doing, spying on a fellow Captain, "...probably won't even find anything... humph!" I muttered to myself in the empty office.

Of course all the 9-5 staff had gone by this time. Within a few minutes I had changed my mind completely. Eureka! There it was in black and pink! Captain X's falsified readings, perfect copies of entries from previous flights. He had painstakingly copied the exact engine parameters from previous Captains – one of whom was ME! I found 6 of them, all showing ECON cruise speed even while the aeroplane was barreling along close to the maximum speed permitted by Boeing, how crazy was this?

What were the odds of these perfect matching engine gauge readings being found usually? I knew that the unions would ask this and it was a fair question, there could just have been a coincidence. I then spent another couple of hours working through all 7 logbooks. There were hundreds of pages... I groaned when I looked at the time, it was dark outside and very late. However I now had clearly demonstrable (mathematical) evidence that the odds were thousands to one against it occurring by accident. Thursday had been a long day I thought to myself as I parked on the drive at my house – kids in bed again and dad not seen for 24 hours. Tomorrow was another office-day for me in the base and I had a pile of other tasks which had backed up this week. I recall at that time in the airline we (the managers) used to laugh and say, "there's always the weekend!" when we were talking about how much stuff hadn't been done during the week.

Friday morning at my desk, I collated all the scanned/highlighted documents into a file on my computer so that they were easily understood. The deception was clearly exposed and he was guilty of falsification of records. Captain X had already reported for duty, he was flying two sectors down to the south of Spain and back to our base in the UK. There was another FO flying with him whom I knew to be a good pilot, but not one of the original three who had made reports already. When they returned that afternoon, I discreetly asked the FO to come and see me regarding another issue as my Rogue Elephant Captain departed for the weekend.

When we were alone in the office, I asked the FO if anything unusual had happened during the flights, to which he replied in the negative. I then explained that there had been other reports from pilots in the base about Captain X and his dangerous flying and that unfortunately it was my duty to investigate such matters.

I asked him if there was anything else he would like to say to me at all about the way the flight had been operated and he said no there wasn't. He actually said "...Capt X operated normally." I couldn't put words in his mouth, I had shown him honestly why it was so important that I should ask these questions and he had still nothing to report. He left for the weekend and I was alone in the office again – I went through all the typed-up statements from the other pilots and thought to myself how strange it was for X to all of a sudden "operate normally..."

It was less than half an hour later when the phone rang, it was the First Officer talking to me from his car, he sounded distressed, his voice breaking up as he said,

"...James, I lied to you and I am so, so sorry!" I held my breath, this was it, here was something very important.

"Something DID happen on our flight back to the UK today. He flies too fast everywhere and we were in a really, really, steep descent in IMC** at very high speed when we had a GPWS Terrain Pull-Up warning! HE DID NOTHING JAMES. He didn't respond at all, I said CAPTAIN! PULL-UP! But no response! He refused to listen to me, but then the warning stopped, we broke out of cloud into the clear and he said 'There it's okay now' – but honestly it wasn't, we could have hit a mountain and he wouldn't have pulled out of that dive..." He agreed to send me his written statement confirming all that he had told me over the phone.

The rest of the story is very sad. I waited for Captain X to enjoy his long weekend with the family (coincidentally mine was wrecked of course by constant phone calls with Engineering – "...what do you MEAN you can't download the FDR? This is a serious flight-safety matter!") The engineers were very helpful in the end and made every effort to track the aircraft from Friday afternoon and secure the recording for us. The only problem was they then 'lost' it (the DFDR recording) for the next 6 days during which I interviewed the suspended Captain X, shown him all the evidence against him and spent a long time listening to his version of the flights.

I also admitted that the engineers had maybe 'lost' the FDR from the Friday afternoon flight – he seemed relieved at this news. Naturally he denied everything, complained that there was a conspiracy against him, including me his Line Manager. He denied emphatically that the crew had experienced any sort of warning on the flight in the descent – effectively calling his FO a liar. In fact all the FOs were liars as far as Captain X was concerned. He even went so far as to visit the ATC radar unit at Manchester Airport to view their tape of the flight when the GPWS warning occurred.

After that visit he stated that the ATC radar tape showed that his aircraft had been descending normally at safe speeds/rates etc. I went two days later to meet the controllers who gave me a copy of the tape showing the opposite – he had lied.

The aircraft had been descending towards the hills at 340 knots (maximum speed) below 10,000 feet and the GPWS hard alert was due to closure rate with terrain.

Finally I had a call from our Eng Dept. They had 'found' the flight-data and were sending it to me, but the file was too large for email – the DHL courier took a day or so while my anxiety mounted. When it arrived, the recording was huge, I had never seen data files so large. I had to scroll through literally thousands of lines of data – the high-speed climb was there, the maximum speed cruise also and then there was the descent...

Starting right up against the barber's pole*** and continuing at maximum speed all the way down until THERE... just below Minimum Safety Altitude at 340 knots... TERRAIN, TERRAIN, PULL-UP, PULL-UP!

*

At his disciplinary hearing with the Chief Pilot, he was summarily dismissed from the airline without payment in lieu of notice for Gross Misconduct. This included 'dangerous flying' and 'deliberate falsification of legal documents/records with intent to deceive'. I felt no satisfaction or joy at this news, simply relief that the Rogue Elephant was gone.

Oddly enough, by coincidence, I discovered that he had left his previous airline (before he joined ours) under very similar circumstances. That is, they investigated him for flying dangerously, low-level at night in an airliner, but on that occasion they had allowed him to resign voluntarily so that his dismissal from the company would not appear on his record. When I found this out, I was deeply upset and spoke with the Chief Pilot from the previous company I told him candidly what I thought... Fortunately there is no record of THAT conversation!

** Rogue Elephant: 1) An elephant living apart from the herd and having savage or destructive tendencies, 2) A person or agency whose activities are antisocial and destructive. (Oxford Dictionary Definition).*

*** IMC – Instrument Meteorological Conditions (in-cloud, without visual references to the world outside)*

**** The Barber's Pole – slang term for the red and white marker on the ASI indicating absolute maximum speed, VMO/MMO. i.e. the speed limit set by the manufacturer.*

*

16 GEAR TROUBLE

By Darragh Owens, Head of Training National Flight Centre Dublin

This bright afternoon I was detailed to fly in a light twin trainer with Dave, a somewhat tentative and "behind the airplane" kind of student. He was working towards his commercial licence. On later reflection, I felt our training organisation might have given him more honest advice, as to whether flying was his true vocation.

But today he was trying his earnest best to get everything right. We had just become airborne in good VFR weather, and he was busy with the after-takeoff checklist. He declared it "complete", with that tone of voice, accompanied by a furtive glance, which instructors will recognise as being a mixture of statement and query. "Negative", I replied, pointing to the one green light still showing on the panel, which indicated the nosewheel remained down. I could tell Dave was mentally kicking himself, for having verbalised the check as complete without noticing the light. I told him to fly below retraction speed while we assessed things.

He checked in the mirror and sure enough, the nosewheel was down. But were the other wheels up? We had definitely heard the gear motors running. Shortly I could see beads of sweat breaking out on his brow. I gave ATC a brief update and requested to fly to a chunk of free airspace. Once established there, I moved the gear selector to "down." No motor noise this time. Still only one green light. No reaction when selecting the switch "up", either. The whole system was kaput. Dave's grip on the wheel tightened, but I said, relax, we have four hours endurance, the weather is good, we'll figure things out. I was trusting my calm tone didn't betray my own level of concern. Landing on one wheel is never a risk-free operation.

I made the mandatory flypast of the tower and Frank, the controller, confirmed we had just the nose wheel down. The next half-hour or so was taken up with Dave flying a gentle holding pattern, while I studied the aircraft manual, tutored him in a bit of structured decision-making, and had a lengthy phone conversation with our chief engineer. We decided the only option was to try the emergency gravity-drop extension. This would be a one-shot deal: if it didn't work then we would be entering a new level of risk and prickly decisions.

In briefing Dave on this, I would see his anxiety levels increasing further still. We set things up at the correct speed and I pulled the lever. There was a resounding thump. Three greens lights blinked on, which was reassuring. We flew past the tower again and Frank said all wheels looked down in the correct position. I told him, okay, I'll fly outbound for a bit and then will turn in for landing.

Given the uncertain state of the gear, I requested crash crews to stand by. That was when it got interesting - because the next thing Frank said was,

"Uh, I have the airport manager here and he's wondering if you could circle for a bit, while we get line traffic in and out." His point was that in their small single runway airport, if my gear collapsed on the runway they would have to divert the incoming regional airliner. Their thinking was quickened by an accident I had been involved in only a month or so earlier, when a student had inadvertently raised the gear on the runway during a touch-and-go. That resulted in us slithering along at speed on the belly of the airplane, sparks flying, propellers bending into corkscrews, and closure of the field for a couple of hours.

The tower's request would have meant us circling in the sky for an hour or more. Dave, trying to be helpful, and sitting at the lower end of the authority gradient, said he wasn't pressed for time. I shot him a reproving glance and said a brusque "Negative" on the radio. There was a pause on the airwaves while consultations in the tower no doubt continued, and then a meeker, "Well, it would be very helpful..." from ATC. I pressed the transmit switch again: "Tower. I. Am. Turning. Inbound. For. Landing." As I explained later, their suggestion was utterly unappealing to me: it would leave me hanging in the air with an uncertain gear deployment, and what would happen if the line traffic itself got stuck on the runway?

Then I would have to travel, wonky gear down, fuel depleted, daylight fading, to an international airport and repeat the whole performance all over again. I would be peeling off our remaining "Swiss Cheese", layer by layer. And if the gear did then finally collapse on a huge jet runway, an even greater mess would ensue.

After a further interval of radio silence, we were given landing clearance. I flew as slowly as permitted so as to reduce our kinetic energy during the imminent encounter with terrain, but in any event the wheels held firm, and we settled on the runway with no further drama. The rescue vehicles returned to base. Dave regained some of his colour, but I suspect the experience put him further off the flying game, at least for a while. He passed his final tests, but I'm not sure if he ever went back to aviation.

It was an interesting one of those "I learned about flying" episodes. My student got to rehearse airborne decision-making in real time, not just in a training scenario. It reinforced the value of taking time, and a measured approach, to analysing problems which occur while aloft. But perhaps the biggest lesson was, don't be afraid to say "No" to air traffic control when their pressing needs, or those of some other operator, would entail reducing your own margins of safety.

17 KICK THE TYRES

By Capt James McBride

Before the flight of every aircraft there is a procedure carried out by the crew. In commercial aviation it is commonly referred to as "the walkround." The function is to act as a final visual inspection of the airframe and engines prior to flight; there will be no further opportunity to inspect the ship. Given the importance of such a task, it is worrying that some pilots appear to treat it so casually. Of course they are aware that the engineering/maintenance team, have already done their final checks on the aircraft.

Even so the principle of duplicate checks is exactly that – in our industry which is safety led, nothing is left to chance. In some airlines (Alitalia for example) it is ordained that the Commander must perform the exterior inspection, which gives a sense of priority. In other airline SOPs it is written that the task 'may be *delegated* to the First Officer' – which again indicates that the final responsibility rests with the Skipper.

This is correct actually because it is the PIC who signs the Techlog (a legal document) to formally accept the aircraft for flight. So, in a two-pilot operation what often occurs is that the PF sets up the cockpit, while the PM carries out the external check including monitoring/ observing fuelling and baggage loading.

I think I have lost count now of the number of FOs who return to the cockpit having carried out the walkround and who either give no report at all, or whose response to my question of "External check okay?" is "Yeah, two wings, two engines, ya'know, the usual..." I would like to think that it's a symptom of modern dispatch reliability and the fact that there are seldom any faults at all with the airliners we operate, but perhaps I was trained in an earlier age. As FOs under training it was ingrained in our psyche that after carrying out the walkround check you always made a short, but formal report to the captain. Words to the effect, "External check is good, wheelwell needs a wash, fuelling completed, aft hold empty." Even now as a matter of courtesy and as part of good CRM, I ensure that I do this myself when returning to the flightdeck for the benefit of the other pilot.

With multi-sector days and a heavy flying roster, I understand there is a tendency towards complacency, especially when you are inspecting the same aircraft (or type) several days in a row without seeing anything unusual. Let us not forget that the operating environment will also play a part. When it is cold and wet outside, maybe it is nighttime – nobody wants to spend too long 'inspecting' the aircraft. However there are some important issues which should be borne in mind, take the Outflow Valve for example – all airliners have one in the hull. Located at the underside aft fuselage, it has electrically operated doors which control an aperture of maybe 30cm x 30cm. It will have driven automatically to the full-open position after the previous flight to depressurize the cabin and therefore that is the state you are expecting to see – both doors wide open. Say for example that it was the first flight of the day and the doors were partially or nearly closed…

This should ring an alarm in your brain because something is definitely wrong here. Think back to the Helios* B737 accident when the crew took off with the pressurisation system in manual – that Outflow Valve was partially closed and MUST have been visible to the pilot doing the walkround. There was a lost opportunity to prevent a catastrophic series of events. Sadly the crew missed other chances too which led to the slow depressurisation of the cabin as they climbed and the ultimate loss of all lives onboard.

Also at the rear of the aircraft, often the cargo door to the aft hold is closed. In the short-haul operation with which I am most familiar currently, the rear hold is normally empty, but if I don't check how will I know? Fortunately with the 737 it is possible to open the door from the ground without special equipment, other types require a belt-loader to climb on or steps. Only by visually inspecting that compartment can you be certain it is empty. There have been instances in the past when airliners have departed with freight in there which should have been offloaded. If not included in the current loading/trim calculation, then you risk an out of trim condition for flight or something much worse.

While on the subject of cargo doors, it is worth paying close attention to the external surface of the door. Pre-existing damage (scratches, slight dents etc.) will already be marked by engineering and therefore will have been recorded in the Dent and Buckle chart/notes in the aircraft Techlog. If there is a fresh scratch or dent this requires further maintenance input to sign off before flight. If your walkround is prior to the first flight with this airframe then at least you are able to say it happened before you signed for the aeroplane!

On some airliners with a long fuselage there is sometimes a tail-skid fitted by the manufacturer underneath the fuselage, to protect the structure of the hull in the event of contact with the runway during takeoff or landing. It is essential that this is included in your inspection, because if the paint is worn away and you can see bright/shiny aluminium there, it is evidence of a previous tailscrape. If you miss this before you take responsibility for the machine, then later on you might have difficulty proving it didn't occur on your flight...

In fact while working for a freight airline in the past I had exactly this situation. Fortunately the hull was blue and the tailskid painted bright red so it was easy to see if any paint was missing. I immediately called maintenance and they sent an engineer.

B767 Tailskid before painting - the paint does not blow away in flight.

I explained the problem, but as his first language wasn't English I thought it better to show him. He came with me to the rear of the aircraft – we looked up at the tailskid (see photo) with its red paint missing and the underside of the skid worn flat with aluminium showing.

He smiled at me and said,

"Ees hokay Captain, Ees normal!" To which I burst out laughing,

"Erm, no it IS NOT NORMAL!" I said patiently, "you can see where the paint has worn away underneath when the skid has been touching the runway doing 150 miles per hour!" He remained unconvinced however and insisted that this was normal. He explained to me that the paint blew off in the wind in flight, which resulted in my uncontrolled mirth, much to his surprise. I couldn't stop laughing.

I responded that I would make an entry in the Techlog now BEFORE I even flew the aircraft and had the photos to prove it. "Engineer must sign it off... AND repaint red paint!" I was adamant. Eventually he conceded and went away to bring another technician to do the tailscrape checks and sign off the Techlog.

On his return with the red paint and the brush (plus the ladders of course) he once again pointed out that it was pointless, because the paint would blow off in flight. Not only that, but he convinced the first officer (another local) to tell me this in English.

"Yes Captain, it will be gone when we arrive in London, the paint blows off in flight, that is what has happened..." I tried to keep a straight face (he was deadly serious) as I said,

"No my friend, paint does not blow off the aircraft in flight or they would all be plain metal tubes we fly, trust me. In the past 30 years I have not seen it happen yet.... But there's always a first time!"

Sure enough, after landing in London Stansted some 5 hours later we both visited the rear of the aircraft together again outside. The newly applied red paint was still intact – the funniest thing was the look on his face. He was stunned! That's the great thing about flying aeroplanes for a living; we never stop learning new stuff.

The moral of the story is do not skimp the external checks. They are a vital layer of defence in our continuous attempts to keep the operation safe. Properly completed walkrounds are an essential part of trapping Errors and managing Threats - we must not just "Kick the tires and light the fires...**"

** Helios Airways flight 522 – 14th August 2005*

*** "Kick the tires and light the fires" – US military pilot slang.*

18 THE LAST FLIGHT OUT OF DUBROVNIK

By Capt Bob Johnson

It was early October 1991 in Gatwick, as I parked my car in the airport staff car park I wondered again why I was here. As an airline pilot, it is not usual to come to work on a day off, but yesterday I received a very direct phone call from the Chief Pilot's PA.

"Peter would like to see you in his office tomorrow morning please Bob. It's a confidential matter..." my reply must have worried her a little because she was eager to reassure me,

"...no, don't worry, it's nothing like that. They have a special flight they would like you to do." Now THAT bit sounded good! Although I was a B757 Line Captain for a charter company and mostly we did charter holiday flights, there was the odd one out.

Maybe it was a juicy corporate flight for a big company – or it could be a football team staying over in Barcelona for a few days... I smiled and decided that whatever it was, I was sure I would enjoy it. I admit I was excited as I came out of the lift onto the 6th floor – this was where all the big wigs hung out.

Almost immediately I bumped into the Senior Training Captain and he greeted me warmly. In response to his question, "...what are you doing in the Ivory Tower?" I simply said,

"I've been called for a meeting in Peter's office, not sure what it's about..." which sounded like I was being evasive, "...but honestly, I haven't got a clue." He winked at me,

"Probably promotion I'd say!" and then launched off down the corridor clutching his armful of papers.

Jane, the Chief Pilot's assistant, was always easy on the eye and looked even more so today as she said,

"Hi Bob, he's in there waiting for you. Glad you could make it at short notice..." I thought to myself, *she must be mid-thirties...? but...*

"Come in!" Was heard faintly through the door after my knock and I pushed the handle. Once inside, I remembered why the Chief

Pilot's voice would be quiet, his desk was the other side of the room and his office was huge. He was not alone. Also, there was his boss, Bill the DFO, (Director Flight Operations) and another man I didn't recognise. It felt odd, they were obviously pleased to see me, almost relieved it seemed... This didn't feel like a Premier Team charter somehow, maybe it was corporate...?

After handshakes all round we sat down. Peter sat behind his desk with his back to the picture window overlooking the main apron, while the DFO sat to one side with the new man. He had been introduced to me as Mr. Smith, a security consultant. To be honest he looked ex-military to me, I've seen enough of them to know. Short hair, clean cut and his posture gave it away. Even sitting in a chair, he was 'Army', or he had been.

Peter began by thanking me for coming in on my day off, which was nice of him and then proceeded to outline that this meeting was confidential.

"Regardless of the outcome, you must not discuss this briefing with anyone Bob, you do understand, don't you?" I nodded and thought to myself, *'hmm ... so it's a briefing...?'* They asked me if I had been following the news about the civil unrest in Yugoslavia and I replied,

"Only insofar as what's broadcast by the BBC, but nothing special. It seems like a terrible mess..." Privately I wished I could appear more informed and make an intelligent summary, but they moved on straight away.

Mr. Smith seemed to know all about it. They let him do the talking now as he started by addressing me formally,

"Well Captain. My job is to monitor such situations on behalf of my company and I am very much up to speed with events in the Balkans. The former Yugoslav army is advancing quite rapidly towards the city of Dubrovnik. As you may know it is an important cultural heritage city with great tourist appeal. The city is not surrounded yet, but we think it will not be long, the airport for example is still open..." At this point the DFO intervened,

"This is why you're here Bob, we have our last tourists still to come home. The Foreign Office have been looking at ways with the Red Cross to evacuate them overland and out through Zagreb, but it's risky and would take time..." Then Peter added,

"We are planning a special charter flight tomorrow night to bring them home. Bob... we would like you to be the captain."

***

As I passed through Peter's outer office again on my way out, Jane was still there,

"All okay then Bob?" she inquired and I wondered how much she knew.

"Yes, fine... everything's fine Jane, erm thanks" She came around from behind her desk and put a small piece of paper in my hand as I was leaving,

"My home number..." she said softly, "give me a call after the flight and we can have dinner together... or something..."

I suppose I could have refused the trip, but the way they put it to me was flattering I guess. They had been through a list of the company's pilots and selected my name and that of an experienced First Officer to come with me. In addition, they had hand-picked the Cabin Crew, we were all volunteers; any one of us could have said no. Our departure time from Gatwick was 8pm.

I later found out that this, the planned last flight of the season, should have been operated two days before, but due to the uncertainty surrounding the security aspects, it had been cancelled. We had 233 seats on the Boeing 757 in charter and no doubt they would all be full coming back. I was informed in the afternoon that we would have just four passengers travelling outbound with us. I was curious; what was this all about? I was at a loss to understand why anyone would want to travel to Dubrovnik at a time like this.

I met Pat the First Officer in the briefing room at the airline's operations centre. It was buzzing with activity. Although there were not many other flights happening, October tends to be a quiet month for charters, it was obvious that our flight was being specially prepared. A new Senior Steward (John) had come over from Luton to take charge of the crew who were mostly male – just two girls out of seven. They all looked focused and serious as I greeted them in the crewroom.

Also, present was the head of Dispatch and the Senior Ops Controller – we didn't usually see these guys after office hours. We got the weather briefing and flightplans. It was straightforward stuff in reality. Just a two-and-a-half-hour flight each way in the 757, what could possibly go wrong?

As the flight outbound was empty (or nearly so) we could take round-trip fuel. Ops informed me there would be no fuel available at Dubrovnik airport anyway and we needed to minimise our time on the ground there. We were hoping to get in and out within 30 minutes. The ground handling company at the airport confirmed they would handle the flight and provide sets of steps and baggage handling for us. There were no tug drivers available, so we would not be pushed back, it would be self-manoeuvring – we would start-up and taxi off the stand ourselves.

They said we would be the only aircraft at the airport anyway. They wanted us to land on Runway 12 and take-off from the other end (Runway 30) to reduce our "exposure to interference" as they put it. That didn't sound nice, but they assured me that Mr. Smith's company had done a very thorough risk assessment. "Tonight, will be okay, but within 24 hours we believe the whole situation will have changed..."

On arrival at the stand I went straight to the fuel truck and ensured the refueller had the correct figure for the ramp fuel. We had decided 26 tonnes should do it. We would have plenty of fuel to hang around airborne for a while on arrival if necessary and then still have enough to get back home. We would be landing at around 75-76 tonnes, well below our max landing weight of 90.

When I saw the loadsheet, I was intrigued to see that our four passengers weighed quite a lot with all their 'baggage'. Then I understood when I saw them boarding. Four very fit guys, dressed casually in jeans and jackets with plenty of soft bags, nothing in the cargo hold. In fact, they looked a bit scruffy really, not at all like military personnel... I smiled as they were welcomed onboard by the cabin crew, they sat in the middle section of a huge empty cabin. As part of the pre-flight brief, I had been informed that these passengers would be leaving the aircraft from the back door as soon as we had stopped on stand at destination – a catering truck would come to the R4 door and they would exit that way. I didn't see them leave the aircraft; in fact, I never saw them again.

Back in the flightdeck, Pat and I ran through the departure briefing and then the pre-flight checklists. The fuelling was completed, and the dispatcher took his copy of the loadsheet and Techlog. We were on our way. I confirmed the passenger figure with the Number One and he closed the flightdeck door quietly. On the overhead panel I could see the door-lights going out as each door was closed by the crew. I looked across at Pat,

"Just a normal ferry flight at the end of the charter season eh...?" he grinned back,

"That's it Skip, just an empty ferry. I'll get the clearance."

You could tell something was different in the controller's voice as she read out our departure clearance on the ground frequency. She was aware there were no flights to Yugoslavia these days. There probably hadn't been any flights to Dubrovnik in the previous fortnight – maybe not since the charter which took our homebound pax out there. We taxied out in virtual silence, as she signed off and handed us over to the Tower, she said,

"Good luck!" Now that was unusual too.

The flight was uneventful as we departed UK airspace, heading for the Adriatic. It seemed like nobody knew there was a war on in Europe, right on the shores of the Mediterranean. Incredible really, but then I suppose it was classed as a 'limited conflict'. It certainly wouldn't feel very limited if you and your family were caught up in it having just gone on holiday to the country. The cabin crew were keen to visit the cockpit more frequently than usual on the way and we were kept well supplied with food and drinks. They confirmed that the four passengers hardly spoke to them and seemed very reserved. Although they accepted the passenger meals and drinks offered, they didn't drink any alcohol, which would have been odd for charter passengers, but we all understood these were 'special'.

Soon enough we were in the descent towards Dubrovnik, talking to Zagreb control. It still seemed normal until we tried to get the latest weather report from the airport. There was no report on the Volmet and no ATIS being broadcast. We started calling Dubrovnik Approach frequency from a very long way away. We got them at around 150 nm out and received the latest info. There was no ILS, no Radar either, but the VOR beacon was still operational and we could make a VOR approach to runway 12. Surface wind was light and variable with only scattered cloud at 3000 feet. They did not make any mention of any fighting near the airport, so no news was good news.

We flew a low drag approach with no lights at all – the cabin was in complete darkness. We had already decided this from much earlier in the evening back at Gatwick – our noise footprint on finals would be close to zero. Regardless of cloud and visibility we just knew it would be safer somehow. We were right. As we passed the ancient city on our left we could see sporadic flashes of light on the outskirts to the south. Although it was dark, with very few lights, there was a pall of smoke drifting off the coast from the city.

It felt weird to be descending towards the airport almost silently in the dark. Although we could see activity on the ground, we knew we would be invisible without lights. The VOR approach was well flown by Pat and I monitored carefully and did the radio. There was only one radio frequency working as we had been informed from Mr. Smith's sources. The controller called himself 'Dubrovnik', nothing else. The runway lights were sparse, some were missing. There was no approach lighting, just the PAPIs and maybe half the runway edge lights – that was all we needed. We switched our landing lights on at a hundred feet on finals and off again immediately after touchdown, we just had the nosegear taxi light.

With Autobrake level 3 and idle reverse, our lightweight 757 slowed down rapidly and without any noise - we didn't wish to advertise our presence to anyone. I took over taxiing as we slowed below 60 knots and Pat took the taxi instructions.

"XXX taxi to the ramp via Delta, there is a marshaller waiting" I don't know what we expected, but the ramp in front of the terminal was completely clear of aeroplanes. There was hardly any lighting as we turned off the runway, we had been briefed to expect stand 10A self-manoeuvring and sure enough a marshaller was waiting for us there, but nothing else. It was like a ghost airport.

When we looked closely there were a couple of helicopters and ugly looking turboprops parked near the cargo buildings to our left at the eastern end, but nothing else. There were no floodlights on the ramp and the terminal looked deserted. There was just an occasional bulb showing from inside. I'd never seen it like this before. It was a shock, but nothing compared to what was to come.

As we shut down the engines, two sets of steps appeared from the shadows next to the building and I also caught sight of a 'High-Loader' catering truck heading our way. We ran through the shutdown checklist and I said to Pat,

"You get the ship ready to fly and I will go and deal with the turnround mate, we need to be quick..." With that I left the flightdeck and headed for the L2 door. The cabin crew had got it open now and the steps were attached, the middle-aged dispatcher was there whom I recognised from previous visits, we shook hands. I noticed he was looking pale and was out of breath, he had been running.

"Captain! Thank you for coming, we have passengers ready, but there is problem..." Of course, after all that we had discussed before our arrival, I wouldn't have expected it to be a 'piece of cake', so I prepared myself for the worst. In a steady voice I said,

"Tell me. What's the problem?"

He said that the passengers had been living in the terminal for the previous two days; I could only imagine what that must have been like. But some more of them had been there for a week since they missed their charter flight back to Bristol with another airline! There had been no flights since. This was bad news indeed. In the background behind him I could see the baggage trucks coming out with belt-loaders too, I mentally breathed a sigh of relief. 'At least I won't need to load the bags tonight'. I was also aware of the smell of smoke from the city, something was burning, and it wasn't nice.

Then softly in the background over the sound of the hubbub of ground operations, I could hear automatic gunfire. Machine guns, with the rattle of other weapons and an occasional explosion. *Shit we really need to go!*

I turned to John and knew immediately the cabin was ready for boarding, I didn't need to ask. Then I turned back to the dispatcher and asked,

"How many are there?" He said,

"Two hundred and sixty-nine Captain..." I winced. Behind me one of the cabin crew let out a groan... I needed to take charge of this situation and do it now. I said,

"Okay, I will come to speak to the passengers. We have only got two hundred and thirty-three seats in the cabin, plus a few spare jumpseats – I think we could take two thirty-seven..." The dispatcher looked at me, opened his mouth and was about to say something, but then just said,

"Come!"

He ran down the steps towards the terminal which was only 30 metres away. I had a hard time keeping up with him, but I caught him at the door to the departure lounge as he opened it. In the distance, the chatter of gunfire continued. There were some lights inside, but not many as I entered the building. Inside was a mess. There were passengers everywhere. All massed close to the departure gate, most had stood up now and they were all looking at us. Parents, children, families, young couples, old couples... here they were, the Great British 'Bucket and Spade Brigade', all looking hopefully at me. He said,

"Here they are Captain" he waved an arm behind him unnecessarily and then gestured towards the microphone at the counter. I know he was expecting me to make a speech about us having, *'only a certain number of seats in our Boeing757 bound for Gatwick and we could only take so many people, we are so sorry....'*

Faintly, I heard a huge explosion, far away in the distance. There was no air conditioning and the smell was bad. People could not be left to live like this. I took a deep breath and said to him,

"We take them ALL. Make sure they have boarding cards, even for Bristol. Fast as you can. GET THEM OUT TO US!" Close by me was a mum and her little girl, they looked desperate. They had heard everything. She caught my eye and simply said,

"Thank you" I nodded, swallowed hard, and headed back to the jet.

As I climbed the steps, John and the other crew members were standing at the L2 door with expectant looks on their faces. Behind me the doors were being opened and passengers were rushing towards our aircraft en-masse. I entered the cool of the cabin, grateful for the Aircon as I spoke to the crew.

"We are going to take them all. There will be around forty odd who don't have seats, it doesn't matter! Get them sitting four to a row where you need to and get 'em onboard quickly we're outta here! OH! AND THANK YOU FOR FLYING AIR XXX!" I grinned at them and they all laughed.

Back in the cockpit, Pat just whistled when I told him and shook his head in disbelief. I explained we would still be ten tonnes under max take-off weight even with all the extra bodies onboard. We ran the performance calculations and could see it was not a problem. We briefed the departure, this time Runway 30, we would head up past the city again and we would use full power for take-off and no derate in the climb.

No lights as soon as we rotated, we would be a roaring shadow heading for the sky. Now the dispatcher was here and offered me a loadsheet, but I had scribbled one out already. It simply showed the flight as being full, there was no doubt about that, I signed it and gave him the copy. As the holds closed, I looked around and our Number One was in the doorway –

"What shall we do about the cabin service Captain?" I replied,

"Don't worry about it, just give the whole bar away!" He then confirmed the headcount with raised eyebrows,

"Okay Skipper, it's two seventy-five, plus four I think..." I nodded in agreement and said,

"Close the doors please John. Let's Go!"

I made a very short PA and thanked everyone for boarding so quickly. I told them it was not a normal flight, but our crew would do their best to keep them comfortable, fed and watered – next stop Gatwick.

I cannot deny it. That take-off was 'A RUSH'! The acceleration was incredible as the full force of our Rolls Royce RB211s pushed us airborne. As I pulled the nosegear off the runway and Pat hit the lights, it was totally dark outside, and I transferred quickly onto instruments. We climbed at Max Angle climb and it was a good ten minutes before we let the cabin crew out of their seats – the deck angle was at least 15 degrees for most of the climb-out.

As we touched down on Runway 26L in Gatwick, there was a roar of applause and cheering from the cabin, we could hear it clearly through the closed flightdeck door. Word must have got around Air Traffic, because as we vacated the runway, the ground controller said over the radio,

"Welcome home boys! YOU DID IT!."

Then there was the reception committee as we parked on stand at the terminal. It was the early hours of the morning, so we were more than surprised to see them all. Our airline 'head honcho' was there, along with other Heads of Department and lots of friendly faces. There was a real party atmosphere as we opened the door onto the airbridge.

PS: Nobody ever asked about that loadsheet or the Pax manifest.

Note: The siege of Dubrovnik lasted for 7 months, 4 weeks and 2 days. Ours really was the last flight out.

*

19 LAUGH WITH THE AVIATORS

By Capt Dan Moore

Right Runway - Wrong Circuit

If you ask any pilot about the very first solo flight, he or she did in the early days of flying school; a significant number of them will have a smile on their faces or burst into laughter. In my case, my face usually turns into a Red-Traffic-Light Sign, caused by a mixture of embarrassment and humour. This still happens to me even though the event took place over 40 years ago!

It was in my early days of the cadetship in Oxford Air Training School in England in 1975, where our National Airline sent us to obtain the British CPL Licence. Of course, English language is not our mother tongue. I was very eager to fly my first solo flight. Some of my classmates got their first solo in a relatively short time like 6 hours, we named them the Aces. Others were extremely slow and got theirs in 19 hours, we named them the Asses. The ones who got theirs in the lower teens, were classified as the Thick ones. I wasn't an Ace of course but surely, I didn't want to be classified as thick, and God forbid, definitely not the other nickname! Luckily my Instructor, John felt that I was ready during my 9th hour of flying.

He pleasantly surprised me as we were doing some circuits in preparation for the big occasion, unbeknown to me, he was planning to give the airplane after the last circuit we did together as a dual training flight.

When we landed on the grass runway and as we were taxiing in, he announced proudly that I was "READY"! That was the last word I heard, my ears switched off completely to the rest of what he was saying, his meaningful words were the equivalent to the Blah, Blah, Blah!

Unfortunately, the control tower changed the visual circuit direction after our last dual circuit. "All Stations, All Stations, it is now a Right- hand circuit", the lady controller was transmitting to all aircraft on tower frequency. All that time we spent in the circuit area, was a Left- hand circuit. Poor John was trying to explain to me the difference between Right and Left-hand Circuits. I pretended that I was listening and understanding what he was instructing me, but I was in seventh heaven, perhaps the same feeling that Neil Armstrong felt when he landed on the Moon. Hey, we are the same calibre, us Aviators and Astronauts!

My confirmation-biased/selective-listening attitude blocked all other meaningful and important information, and singled-out only one sentence, "READY TO FLY SOLO NOW." I remember nodding yes to him seconds before he firmly shut the Piper Cherokee 28 door saying, "Dan, it is a Right-hand circuit now, you turn right to join downwind, understood?" He tapped his fingers on the canopy, thumbs up and waved bye-bye to me. I was on a different planet than he was, saying to myself, "what is wrong with John, does he doubt my ability and what is this turn *right* he said? We were turning left all the times and worked for me." I held for a few seconds to check the magnetos, all good. Contacted ground asking for taxi clearance; "Oxford ground Uniform Golf requesting taxi clearance."

"Uniform Golf clear taxi to the holding point, for a RIGHT-HAND CIRCUIT", the controller replied with a loud voice, deliberately emphasizing the right-hand circuit. What I did not realise was that when John left the aircraft he went running to the control tower which was only metres away where I was doing my pre-departure checks. I acknowledged "UG." As I was taxiing out, I was waving happily and proudly to a lot of my classmates who had gathered at the fence isolating the flying area from the students' pub some twenty metres away from the taxiway. Then I was instructed by ground to contact the tower frequency. "Uniform Golf, cleared lineup and hold", was the term then. I acknowledged, "Uniform Golf."

"Uniform Golf, it is right-hand circuits now, acknowledge please", replied the tower.

"Uniform Golf", I replied confidently while rolling my eyes and saying to myself, "what is wrong with those idiots?"

"Uniform Golf clear takeoff runway... wind is calm, Right-hand circuit."

"Clear take-off Uniform Golf", I acknowledged and off I went.

As the nose lifted off the ground, so was my spirit. First time ever, my first solo flight in a powered airplane! Incredible feeling. "Liftoff we have, Liftoff!", mimicking the cliché of rocket launching into space. Shortly after, I was steering the Piper to the left. The lady controller started screaming, "Its Right-hand circuit, RIGHT-HAND UNIFORM GOLF."

As if I had everything under control, I replied with confidence, "Uniform

Golf", while continuing the left turn to join left pattern downwind. I hate to imagine what the controller was going through then. Suddenly I heard, "All stations, ALL STATIONS! Student FIRST SOLO joining left downwind." I was elated and delighted that she was announcing to everybody on the radio I was on my first solo and after only 9 hours training at that!? I chose to ignore all other instructions (had no idea what was at stake here), so I said to myself, "I better acknowledge her call, she seemed frightened." "Uniform Golf."

The 4-seater Piper PA28 - a popular training aircraft.

There were about three aircraft in the circuit in various positions on the Right-hand circuit. Other aircraft were rejoining the circuit now from the flying training area. The controller was giving instructions delaying some from rejoining, gave others to maintain altitude, and giving others instructions to hold. As I was turning left base, I noticed two aircraft, one of them was head-on coming from a right downwind and one more on a late right downwind. I said to myself, "what are those idiots doing." I had the audacity to tell the tower, "There are two aircraft on the wrong downwind."

It was hilarious to hear the loud laughter of the controller, she was replying while she was choking and could hardly speak.

"Affir..ma..tive, Uniform Golf, you are correct. YOU are number one to land, Wi..nd.. is, is.. calm.. you... are.. clea..r to La..nd, BREAK, BREAK. Zulu Hotel, climb immediately to 2000 feet, Mike Delta, extend your long downwind."

Well, I had done my duty, I had informed the tower of the wrong-doing of the other aircraft in the circuit. I was very proud of myself and my performance. Shortly after, I lowered the flaps to full position and here I was on short final. Of course, what mattered then for the students was how smooth was the landing? If there was no bouncing, you got the full mark. Well, mine was a greaser, or at least that's what I thought.

Back to the apron, shutdown the engine, parked very close to the tower and I saw my Instructor John with two other instructors and a bearded man, the ATC Duty Officer coming toward me with big smiles. I thought that I must have impressed them with my landing. There was no blame, no scolding and I was escorted to the principal's office by John. An office we dreaded, a student goes there either to terminate his or her training or be awarded. Until that moment, I thought I was going to be rewarded.

Luckily the Principal was in a good mood, sipping his coffee while his back was to us and he was facing the window overlooking the airfield. He turned his face towards us and asked,

"Dan, do you know why you are here?."

"No, Sir, but I think you liked my landing..." He was shocked, to the degree that he spat out the last sip of coffee he had and burst into laughter.

"Son. You need to listen and understand instructions, John will talk you through what happened, but you are in luck today, we are not going to terminate your training."

The rest is history.

Dan became a commercial airline pilot and is now an airline captain in the Middle East. He never forgot this early lesson on confirmation bias. He is a pretty good listener these days.

*

20 ICED UP – GOING DOWN!

By Capt James McBride

"London Control, Grosvenor Zero Three Bravo, request descent!" my voice on the R/T may have sounded calm, but I felt far from it. The reassuring voice in my headphones returned with.

"Standby Grosvenor Zero Three Bravo, maintain Flight Level Niner Zero..."

My situation was not good and I knew it. At cruising altitude, nearly nine thousand feet above the Pennines in solid clag. I looked again at the Airspeed Indicator, *hmmm... nearly back to blue-line speed** and I had full power on... A quick scan of the vertical speed indicator showed the needle dipping slowly, the aircraft was just beginning to lose altitude and I knew what was coming next. I twisted slightly in my seat and shone my torch through the side window of the little cockpit – *Ugh!* I mentally winced – the air intake of the left engine was a mass of ice and snow, clogged.

I didn't need to look at the right engine, it would be the same. I could see the horizontal snow streaking past the glass outside, a blizzard. Also I was aware of the irregular banging noise on the fuselage as ice was being shed from the props - the electric anti-ice heaters were doing their job. I had just tried to de-ice the leading edges of the wings again with the inflatable rubber boots, but the effect was negligible - the ice build-up was fast and furious. All of this took only seconds, before I decided I couldn't wait any longer.

"London. Grosvenor Zero Three Bravo *descending*... leaving level nine zero, erm... looking for lower..." how much lower I didn't know, but I knew it was impossible to stay up here. *Maybe I could get below the freezing layer and some of this damned ice would melt...* fat chance of that I thought miserably. From the forecast it was freezing on the hills tonight over the whole route. Damn! I can't let it go any slower, *must maintain the airspeed above... the... blue... line.* I trimmed again so she would fly hands off. A Weather Radar would be nice I mused.

Oh to have a Weather Radar, but this ship is too small and too cheap for that, the R/T came to life again.

"Grosvenor Zero Three Bravo, London. Sorry about that. Was on the phone, are you unable to maintain altitude? Be aware you have loud background noise on your transmissions."

"Yes London, THAT'S THE ICE! Against the hull – coming off the props. Erm... We've picked up a load of ice here and full power on, *we're going down...* Unable to maintain cruise level and request a left turn... ah, sixty degrees, maybe we can get clear of the weather?" it was a long transmission and I knew it, but it was a big story for me. I thought, *why do I keep saying "We", there's only me up here....!?* He was quick to reply.

"Turn Left, heading 280 degrees Grosvenor, are you declaring an emergency?..."

It is a well-known fact that Ice and aviation don't mix well. In reality it is worse than that, Ice is a killer. The pages of aviation history books are littered with the reports of hull losses due to icing. Airframe ice and/or engine ice, during winter operations the dangers are serious and should not be underestimated. If the atmospheric conditions are right, ice will accumulate on cold airframes and in engine intakes with adverse effects on aircraft performance. An example comes immediately to mind; Air Florida Flight 90 in 1982 - a B737 which stalled after takeoff and crashed into the Potomac River in Washington, killing 74.

Back in the late 1980s I worked as an Air Taxi pilot at Manchester for a couple of years, just prior to joining the airlines. It was varied and interesting work and in those days we operated all of our flights as single crew. This was in the days before regulations made two crew Ops mandatory or solo with a working autopilot. Not only did we fly alone, but we also had to fly the aircraft manually on instruments. The Partenavia PN68s were light piston engined twins with propellers and as we flew in all weather in the UK, we knew all about icing. There was one regular mission which we used to fly and I remember it well. The company had the contract to fly from Heathrow to Prestwick every Wednesday night. It was technically a freight flight, but the cargo itself was not heavy – it was the printing plates for the

Sunday (and Radio) Times magazine. Bear in mind this was in the days before the internet and the silicon chip. The originals were made in London and as the media organization were time sensitive - always up against a deadline - they had to ship to Scotland on the Wednesday night without fail. A motorcycle courier would bring the 3-kilogramme package to Heathrow private jet terminal where we would be waiting.

Partenavia PN68B - affectionately known as "The Prat."

During the summer months, I recall this flight being very popular with all the pilots – we positioned empty from Manchester to Heathrow and then after delivery in Prestwick, empty back to base. It was a doddle. Wintertime was another story, nobody volunteered and we all had to take our turn. October 1987, it was my turn – unlucky me. Flying unpressurised propeller aeroplanes the length of the UK below 10,000 feet at night in IMC was not funny. One saving grace was that we were operating at light weights, the other was that it improved our instrument flying skills, we didn't blink much.

The icing protection systems which we had on the light twins consisted of rubber boots on the leading edges of the wings and electric heating elements in the propeller blades. In icing conditions, the prop de-ice was switched on routinely of course as an anti-ice measure, but the wing de-ice boots were different. As ice built up on the leading edges of the wings it was important not to use the de-icing boots too early. The reason was that if there was only a light coating of ice, then under certain conditions it was possible for the rubber boots to expand but not break the ice away.

This would leave an enlarged frozen bubble on the outside of which more ice would form. The danger then would be that there could be a huge accumulation of ice which was impossible to remove... On this particular night flight, although I did not believe I had made this mistake, I was certainly in a bad situation where the airframe was picking up more ice than it could cope with. The increased drag from the ice and reduced power from the engines combined to have such a negative effect on performance that keeping level flight in these conditions was not possible, I/we had to get out of cloud.

Am I declaring an emergency? I thought for about half a second... *yes good idea!* It was weird. Part of me felt guilty for causing a nuisance, after all I should have been able to maintain altitude. I was shocked that it was not possible, it had never happened before. These days in Human Factors research it's called the Startle Effect, but in those days we didn't know so much.

"Affirm London a Pan call please, Pan-Pan Grosvenor Three Bravo descending, unable to maintain Cruise Altitude due to ice..." now steady on the heading, *surely we can get out of this.*

"Roger Grosvenor Three Bravo! The base of the airway is flight level six zero and you have no traffic below. *Cleared descent."*

I wanted to hug him. I focused on flying the dials. Slowly rolling into the turn, passing 8,300 on the altimeter, VSI nearly 500 feet per minute down and still at blue-line on the airspeed indicator... The controller came through again,

"And the Regional pressure setting is Niner Eight Zero Millibars." Very kind of him. I would need this as I dropped out of the bottom of the airway to judge my height above terrain. *TERRAIN...!! Argghhh,* the thought had me looking again at the safety altitude on the chart – hmmm, present position, just over four thousand feet. 980mbs, a typical low QNH in the winter, I set it on the standby altimeter – approaching 7,000 feet at this time.

A quick glance at the engine gauges showed maximum power with alternate air selected. Of course 'ALT AIR' reduced the power output, but as the normal air intakes were blocked solid, there was no choice. I recalled stories about pilots from the old days playing with the engine fuel mixtures separately on each motor to make them backfire to blast the ice out of the intakes. *Jeez, I hope it doesn't come to that,* (Rising panic...) *I DON'T EVEN KNOW HOW TO DO IT... Can't risk an engine failure here.*

Suddenly the horizontal snow stopped snowing – I breathed heavily, unaware that I had been holding my breath. We're showing warmer air on the OAT gauge... then all of a sudden, CLEAR! We are in the CLEAR! Out of cloud, I can see the lights of a city below, Yorkshire I am sure - Leeds, maybe. The snow has definitely stopped and I press the de-ice boots switch again. Looking hard with my torch along the leading edge of the left-wing I see the ice breaking away as the rubber boots fill with high pressure air – YES! And I can FEEL the ship start to perform better now, airspeed needle rising and we are flying level. No descent on the VSI, phew! My beautiful little aeroplane...! Come on Hun, you can do it girl...! COME ON!!

"LONDON, WE'RE CLIMBING AGAIN!" I need to tell somebody, he replied immediately,

"Roger Grosvenor Zero Three Bravo, maintain Flight Level Seven Zero - call Scottish Control now on frequency one two six decimal three, good luck."

I thanked him sincerely, cancelled the Pan and changed the radio over. I noted that we were going faster and faster now. Soon there would be a surfeit of power, luxury. I was relieved for sure, checking the weather again I could see we should be in the clear all the way to Prestwick. I looked forward to a celebratory cup of hot coffee! Oh the joys of the Air Taxi pilot in the winter.

** "Blue-line speed" = Best rate-of-climb speed for a twin-engine aircraft with one engine inoperative. (Also known as safe single-engine speed).*

*

21 AVIATE, NAVIGATE, ADVOCATE

By Capt James McBride

There is a concerted effort in modern CRM teaching these days to ensure that inexperienced first officers are aware of their duty to speak up during line operations in order that they contribute effectively to the decision-making process. In several companies, this is referred to as encouraging F/Os to 'Advocate' their thoughts and opinions when faced with situations requiring the cockpit crew to reach a decision.

You may wonder why this should be an issue. It is because history has proven that many airline accidents occurred where there was a steep authority gradient in the flightdeck between a senior, experienced Commander while operating with a junior colleague. In the final analyses of these events, if the junior F/O had been effective in putting their view across to the captain, then many would not have occurred.

In July 2010, an Airbus A321, operating flight 202 for Airblue airlines crashed into the Margala Hills near the destination airport Islamabad - the capital city of Pakistan. The flight was coming in from Karachi and was making an instrument approach in marginal weather. It is significant that a flight by China Southern Airlines had diverted only 30 minutes earlier having failed to gain visual contact with the runway. The bad weather at the time of the accident included poor visibility, low cloud and heavy rain. The accident investigation subsequently established that the A321 (AP-BJB) had been serviceable at the time of the accident – so this was classified as a CFIT* event.

Early reports from officials were quoted by the BBC "there was nothing in conversations between the pilot and the Islamabad control tower that suggests anything was wrong." However the CVR recording when released by the investigation team later was at odds with this statement. In addition to the ATC tower instructions/advice to the captain to turn away from the high ground and the EGPWS** aural warnings of "TERRAIN AHEAD", there was also the First Officer's voice which was heard saying; "Sir turn left, Pull Up Sir. Sir Pull Up..."

By this time it was apparent that the 61-year-old aircraft commander with over 25,000 flight hours in his logbook had become disoriented and was displaying lack of situational awareness. The resulting crash into the hills cost the lives of all 146 passengers and crew.

The chain of events leading to the aircraft being flown unintentionally into high ground had begun a lot earlier in the flight however. The investigators learned from the CVR that for the first hour of the flight, the captain made a point of testing the First Officer's technical knowledge of the Airbus and for the rest of the flight the F/O became very quiet. Point 10.5 from Chapter 10 of the official report by PCAA states:

"The captain's behaviour towards the FO was harsh, snobbish and contrary to establish norms. This undesired activity of the captain curbed the initiative of the FO, created a tense and undesirable environment, and a very conspicuous communication barrier in the cockpit, portraying a classic CRM failure."

Significantly later on in the flight, when the captain decided to fly a Circling Approach to the runway, he instructed the F/O to insert a series of waypoints into the Flight Management System which he will have known was unauthorised and illegal, but he did not challenge the captain's decision making. Chapter 11 (Conclusion) of the PCAA report includes the specific point 11.6 which reads:

"FO simply remained a passive bystander in the cockpit and did not participate as an effective team member failing to supplement/compliment or to correct the errors of his captain assertively in line with the teachings of CRM due to Captain's behavior in the flight."

The report also included the alarming fact that the captain ignored 21 repeated aural EGPWS Terrain and Pull-Up warnings over a period of 70 seconds before the aircraft impacted the ground. While the report of Flight 202 and the breakdown of CRM was singularly shocking, sadly it was not unique. In the last decade there have been many hull-loss accidents which have been a result of similar human factors failures. Which explains the current and renewed focus on CRM training in order that inexperienced first officers are not shy to

'Advocate' their views.

The caveat to all of this of course is that the FO must know what they are talking about in the first instance. I have experienced some interesting situations in the not-too-distant past while flying airliners when the FO has voiced his/her view without having the in-depth technical knowledge to back it up. A case in point was when we were at high speed, shortly before V1 taking off from an airport in London bound for Morocco.

I was the captain, flying as Pilot Monitoring and the first officer was PF. The aircraft was a Boeing 737NG. The Master Caution lights illuminated due to a Thrust Reverser light in the overhead panel. There were no other fault indications, no accompanying airframe or engine vibrations, the engine parameters appeared normal and so I stated,

"Continue" in accordance with SOPs. Very soon afterwards I called "Vee One – Rotate!" and we got airborne.

Passing 400 feet the First Officer asked me to,

"Restate the Failure," and as I looked at the overhead panel with the glaring REVERSER light brightly illuminated, I did have a strong desire to say,

"Mate – you really don't want to know..."

I resisted my natural urges however and simply said what I saw

was the problem.

"We have an indication REVERSER light on the number two engine," to which he correctly responded,

"No memory items, we will action the QRH drill once the after-take-off checklist is completed." We did this and considered the problem and our options. As my young colleague had been well trained in a modern CRM environment with our airline, I was comforted in the knowledge that I could rely upon him for positive input to help us resolve the situation. What I was not expecting though was his immediate statement...

"WE WILL HAVE TO GO BACK!" He was quite adamant about this until I said to him,

"Steady on chap, let's not jump to conclusions. We are six tonnes over max landing weight for a start..."

Of course the Maximum Landing Weight does not apply in an emergency situation, but this was a non-normal situation and not in the emergency category. In accordance with the QRH drill for the aircraft which we had both read and actioned together if the Reverser light was illuminated without any other symptoms, then we could continue the flight to destination.

We did not have to go back. So we discussed it further, I called the Airline Engineers and Ops departments who agreed that we could safely continue the flight. In fact they very helpfully suggested that we land at a maintenance base enroute to Morocco to have the Engineers lock-out the #2 Engine Thrust Reverser so we would not have an MEL/No-Go item on the ground in Morocco. We coordinated with the cabin crew; informed the passengers; arranged for the Maintenance Team to meet us in Southern Spain and then planned our approach and slightly overweight landing at our enroute alternate airport. All in all quite a busy day in the office. The actual turnround was very efficiently carried out and the handling agent even met us on arrival with our next flight plan as we opened the doors.

My abiding memory of it all though was that the F/O's immediate response was "WE WILL HAVE TO GO BACK...!" without properly considering all of the implications and consequences. Although it is unusual (and somewhat disquieting) to see a Thrust Reverser light illuminated inflight with the worry that the reverser might deploy - I explained that there were several interlocks which would have to simultaneously fail for that to happen.

In fact I assured him that a few years ago, I had personally flown a 737 from the Middle East through the night to the UK (refuelling on the way) with an identical fault. This was true, but THAT as they say is another story...

The moral of this story is that while the modern CRM trends are welcomed, inexperienced aviators must remember that the old goat sat in the left-hand seat has usually seen an awful lot of stuff before. Yes by all means 'Advocate' however this should come after Aviate and Navigate in the time-honoured tradition. Part of 'Aviate' is being entirely familiar with the contents of the QRH and technical matter concerning possible system malfunctions. Don't start off advocating with "WE WILL HAVE TO GO BACK!"

Most important of all, try not to jump through the decision-making process without communication between not only the two pilots onboard, but other outside agencies. The Engineering Department on that day were extremely helpful to us and gave excellent support, which enabled the flight to continue with a minimum of delay in safety. What more could we ask for?

*

22 PETER'S STORY

By Capt Nick Francis

Peter set his glass down on the flying club bar. Turning to look at me, he asked,
"Do you believe in ghosts?"
"Yes" I replied, "never seen one though."
"Neither have I, but I smelt one once" he said.
"Smelt one? How can you smell one?" I was intrigued.
"It's an interesting story, but strange," he answered. "It's strange for three reasons. The first is that as an aviation ghost story it happens on the ground. Secondly it occurred primarily by sense of smell, and for the third. Umm, well I'll tell you that at the end"
"Huh, I suppose this round's on me then if you're going to start telling great yarns."

I refilled our glasses and we settled into the worn leather armchairs that had graced the clubhouse since we had started our flying careers there many years before. Peter took a sip of his beer, and settling his gaze onto the large pane windows looking to the west he began his story.

"I have to take you back twenty years. I had not long got my first jet job. You know which company. Having completed the very tough conversion syllabus that they applied to the Boeing 737-200 I knew very well that the six-monthly simulator checks were not going to be easy. The company prided itself on outdoing its competitors on training standards. As the date of one simulator check approached I sank to a deep low when I saw that I had been rostered with a very zealous trainer and a captain to whom the finer points of good 'cockpit resource management' would not become apparent this century. Indeed it was still the era of the 'man and a boy' operation, when 'men were men' and captains were still gods.

The simulator we used was that of Aer Lingus in Dublin. As 737-200 Sim's went it was quite good and generally felt like the airplane. It was housed in a bright, modern, purpose-built building along with the other Aer Lingus types operated at the time.

As you may remember, Aer Lingus placed the name of a patron saint on the nose of all their aircraft. So it was that someone with typical Irish humour painted 'St Thetic' on the nose of all the simulators.

This always succeeded in raising a smile from those of us treading those final few yards to the torture chambers assigned to us for the coming four hours. Indeed I have often wondered as I walked down through the clinical doors and corridors, to be faced by the gantry stretching over to that dreaded metal box if any similarity existed in the emotions felt between that walk to the simulator and the final walk to one's fate in an American prison!

The first day came, and with typical zeal the trainer manipulated the scenarios to his advantage. The captain and I sweated profusely to cope with the multiple failures and poor weather given to us. With the captain operating life was simple and all I had to do was his bidding, read the checklists, and not answer back.

With me as operating pilot however the pressure was on. The company was a new one and had recruited all its first officers with a view to early promotion. Thus every Sim check mattered as one was expected to start one's command course within a year or two of joining. My performance that day went from bad to worse as the pre-check nerves set about undoing my best intentions. Decision making was slow and irrational and I began to over- control more and more. The 737-200 is not forgiving of such actions. It was obvious that the trainer was becoming increasingly impatient at my failings, whilst my Neanderthal colleague in the left-hand seat sat in smug satisfaction that his previous low opinion of me was completely justified.

Thus the first day's session ended and after debriefing we adjourned to the airport hotel. This was an old hotel next to the airport which had been used for many years by airlines to overnight their crews. After settling into my room I pondered on the thorny question of whether to jump off the roof before or after dinner! However having always been partial to a pint of Dublin's favourite beer, and not being one for ignoring an empty tummy I opted temporarily for the latter.

Suitably filled with both food and 'refined Liffy Water' I retired to bed to ponder my fate on the morrow. My nerves were keeping me awake and to be honest I felt that if I performed in the same manner tomorrow as I did today, then I would be at risk of not only failing promotion, but the very renewal check itself. Thus I lay in bed staring at the ceiling, wide awake.

Now the room in which I lay was air-conditioned with a closed window. No apparent odours of any kind were apparent on my entering or retiring to bed. But as I lay there I became aware of the smell. To describe it would be difficult, 'Oil and Leather' would be close, but to those enthusiasts of vintage aircraft... The unmistakable smell of an old flight deck.

I have only known such a smell when sitting in an old aircraft or car. It's a wonderful odour, evocative of a bygone era and the characters that plied their trade in an environment that called for skills now long gone." I nodded in agreement and took a very large swig of beer, conscious that the hairs on the back of my neck were beginning to stand up,"

"Go on..." I said and he continued.

"At first the smell was slight, barely noticeable, but slowly and steadily it grew stronger, becoming so powerful at its peak that one could be forgiven for thinking that one was indeed sitting in the flight deck of a DC3 or a Bristol Britannia. At first I was just curious as to the source, but then I became aware of a second, and far more serious sensation... I was NOT alone!

Oh, I don't mean some form was hovering over the bed with ghoulish face or that the bedsheets had arisen and developed eyes. No, when I opened my eyes there was nothing there at all. But I was NOT ALONE! I knew it. I was as sure of it as I am of being here now. Something or someone, was in the room with me. I felt no fear, just enormous curiosity. That is not to say that I was being brave. No, somehow I knew that I had nothing to be afraid of. Whoever my visitor was, he or she meant me no harm.

I have always believed in ghosts and have long relished the thought of seeing one. So I was almost beginning to feel cheated at only getting the redolent version when the purpose of the event became clear.

Gradually I became aware that my mood was changing. Not by choice, but by manipulation. My visitor was trying to tell me something. Not through little voices, but by a deeper level of subconscious communication. The message was very simple "Relax, go to sleep, and all will be well tomorrow" and I found myself willingly succumbing to the message. It was as if a warm wave of relaxing inner peace was washing over me. It was a wonderful sensation.

So, enveloped by my new-felt confidence I thanked my visitor and wished him or her a good night, and rolling over, I did what I was bid. I fell into a deep and relaxing sleep.

The next morning I pondered on my visitation. Had I imagined it? No. It was too real and certainly happened before any drowsiness had overcome me. Had my worried state engineered an event in my mind to counter the stress I was under? Maybe, but I preferred to think that some caring aviator of a generation far long gone had taken the time and trouble to give me the help I needed with my predicament.

Although I will never know who or why. So did I go to the simulator and fly the pants off it? Oh no, my story, strange as it may be, now takes on the most incredulous element of all."

He paused and we took another sip of our beers, he had my full attention.

"The second day started as bad as the first. Captain Neanderthal barked his orders. The trainer invented ever more complex multiple failures and manipulated the weather to his advantage until as they say, we were 'hung out to dry'. The crux of the afternoon came as I was carrying out a single engined NDB approach at Birmingham. To make it 'more sporting' the trainer decided to throw in some thunderstorms with moderate to severe turbulence. Thus with the windows flashing to simulate the lightning and the sound effects emulating the thunder the simulator kicked and bucked in my hands as I struggled my way around the procedure.

If I remember correctly it was on the outbound leg that it happened, just after I became aware of a much more realistic clap of thunder. All of a sudden the column went limp in my hands, we lurched in our seatbelts and the simulator collapsed onto its side with a resounding 'thud' as it came to rest on its jacks.

At the same time every single warning and caution light came on and every bell, clacker and siren let loose at full volume. As we took in the shock of our predicament a smell of electrical burning was also becoming evident. I looked quizzically across at Captain Neanderthal, who for the first time passed a sympathetic look my way. Smiling he said,

"I don't think there's a checklist for this one Peter, let's get out of here."

As we turned away from each other to gather our belongings I smelt it. Yes, there it was, quite unmistakable, and getting stronger over the electrical fumes. 'Oil and leather', just as clearly as I had experienced it in my room the night before.

I turned to my left and asked,

"Do you smell that"?

"Smell what?" he replied.

"Oil and leather, old flight decks, you know... a vintage aircraft smell."

"All I can smell is us frying if we don't shift ourselves out of here soon," was his reply. And joining the trainer who had already leapt the gap to safety, we vacated the simulator and set off for the debriefing room.

It transpired that the trainer had perhaps engaged a little more realism than he bargained for when he set up the thunderstorms. For Dublin was experiencing the very same that afternoon and the simulator building had been struck by lightning. That was the cause of the computer crash and the electrical smell thereafter. Of course you may draw your own conclusions as to the smell of the old flight deck.

As it was going to take some time to reload the computer and our booking would expire, we cancelled the detail and headed home to our base. With my check incomplete of course I had to return to the simulator soon after, but this time with a trainer that I could relax with and a Captain that I enjoyed flying with. The check went well, and soon after I started my command course, passing it later that year.

To be fair, the first trainer became a strong and supportive friend during my course. Perhaps he reckoned I had come through whatever his personal requirements were to make the grade. As for Captain Neanderthal? I never flew with him again and I believe he has long since retired."

Peter settled back into his chair, and sipping his beer looked at me for a reaction. I said,

"Well Pete, that was different. How long did it take you make that one up?" to which he smiled.

"Oh I didn't have to. I said I'd tell you the third reason why my story was strange at the end. It's very simple really. It's strange because it's absolutely true!"

I have not seen Peter since. We keep in touch but our airlines operate at opposite corners of the world. However, I still get to operate through Dublin occasionally and have often stayed in that same old airport hotel. Always before I fall asleep I think of Pete's story, and whether 'St Thetic' as I have come to call him has helped other frightened and worried pilots in the course of their careers.

I do hope so.

(Editor's Note: I puzzled over the meaning of "St. Thetic" for a minute or two and then realised it is a play on words. Saint-Thetic, is very similar to "Synthetic" which is a description of the Flight Simulation Training Devices used to train airline pilots. Often it is called Synthetic Flight Training).

"Do you smell that? Oil and leather, old flightdecks you know, a vintage aircraft smell?

"All I smell is us frying if we don't shift ourselves out of here…"

23 WINTER OPS IN JULY...

By Capt James McBride

The last of the passengers were boarding and the cabin crew were busy in the cabin of our Boeing 767 helping them to stow baggage and find their seats. As it was a military trooping flight, there was rather a lot of 'baggage'. There was cargo too. The aircraft holds were pretty much full of essential freight for the military base - our destination airport. In the flightdeck we were completing the pre-flight preparations. I looked again at the Weather and Notams for destination. 'Moderate Snow' was falling and forecast to do so for the next few hours, after which it was forecast to get better. Additionally the destination NOTAM stated 'SNOCLO'... Airport CLOSED due to snow!

Under normal circumstances, the dispatch of this particular flight would qualify as insanity, but these were far from normal circumstances. The schedule had already been delayed for more than 48 hours and everyone was pushing hard for us to go. There were military personnel waiting to be transported out of the base after finishing long tours of duty and two stretcher cases needed to be CASEVAC'd to the UK in the portable hospital section of our aircraft. Plus we were carrying much needed supplies and replacement people into the Theatre of Operations.

So the pressure was on. Not only that, but our major enemy at this time was Mother Nature who had intervened with a succession of Cold Fronts dumping vast quantities of snow on the airfield. Of course you might question why we should be talking about Winter Operations in July...? Well the reason for that is because our destination was Mount Pleasant Airport in the South Atlantic, Falkland Islands. Remember when we British kids learned at school that "Aussies eat their Christmas Dinner on the beach"? The same logic applies, everything is upside down in the southern hemisphere and July is deep winter for them all. Naturally we had prepared well for this flight – we'd certainly had plenty of time while kicking our heels for a few days on Ascension Island just south of the equator and over 1,000 miles from Africa. The inbound flight from UK had arrived 2 days before and we, as the new crew were ready to launch south immediately, but the weather put paid to that.

There had been a series of snowstorms blowing through and although the military had cleared the runway a couple of times, it was to no avail. One of the problems was that once they cleared the runway and declared the airport open, it was too late to launch from Ascension Island (ASI) because the flight time was eight and a half hours. We could see from the TAF* that more heavy snow was imminent and could predict the runway would be blocked by the time we got there – and so it proved.

Naturally everyone had been pretty frustrated by this sequence of events, so we had to adopt a new strategy. We liaised with the senior Met people at both ends of the route and we could see there was a weather window coming up. It was not going to be huge, but should be enough for us to land, provided that the Army and RAF cleared that runway for us and kept it clear...

The downside was that our departure time from ASI was the middle of the night and we would fly through the night to arrive shortly after dawn. We reported for duty at 2100 local time to have our brief with the on-duty Met officer. She was a nice lady but had not been involved in our discussions previously and only just returned from leave. When she saw the weather forecast and the NOTAMs (saying SNOCLO) she informed us that we should not fly.

I took some time to explain to her that we had been through all of the scenarios and although the forecast was bad at the moment, it was a flipping sight better than what had gone before – she was not convinced. Outside the aircraft was being fuelled up and passengers and freight were being loaded – now was not the time for a row with the on-duty Met officer. I said in a much calmer voice than I felt inside.

"Okay, you've provided the up-to-date forecast, which is similar to what we planned for 36 hours ago. We are going to depart now and have been assured by the military at Mount Pleasant the runway will be clear for us. Thank you for the Met Info, but as Aircraft Commander I take the responsibility if it all goes wrong and we divert the flight, nobody will blame you."

She seemed content with my assurance – well she looked happier than I felt anyway. At the back of my mind I knew that a diversion would be a major problem, because the only usable alternate airport was Montevideo in Uruguay! The flight time if we had to divert there was 3 hours or more – we would have to fly off the coast of Argentina all the way north. Carrying enough fuel for this mission would put Boeing's "ER" suffix to the test – we would have full tanks for takeoff. On the plus side, we would have enough gas for about one hour holding over the Falklands waiting for the snow clearance teams to do their stuff if required.

I had been able to get through to the Senior ATC Officer on the phone earlier in the day and he assured me that the snow clearance teams would have the runway open for our arrival, *provided the snow stopped falling* when the met office predicted... That would be a few hours after our departure, but by that time we would be out of radio contact somewhere over the South Atlantic Ocean – it's a really big place. In fact I was not even certain that the guy I spoke to was actually the Senior ATCO, "comms" were not easy and it had taken quite an effort for me to get through on the phone at all! And this from the 'communications island' which received the first words from the moon landing back in 1969 by Neil Armstrong, "One small step for man..." etc.

Later with all passengers boarded and our checks complete, they were closing the cargo doors - we were all set for departure. In the dim floodlights of the flightdeck I looked across at my young colleague, "Call for start-up please" and now on the headset came the cheerful voice of the groundcrew chief Spike, "All doors and hatches closed skipper, chocks in position we're ready down here..."

"Thanks Chief, the parking brake is set, Standby..." – another look across the cockpit and it was clear that the First Officer was having some sort of discourse with ATC, *what NOW?!*

In response to my questioning looks he said,
"The Met Office lady wants to talk with us on the radio..." I replied,
"Okay, Ops frequency on Box Two - Let's see what she wants."

Her voice was urgent and concerned as she transmitted from the Ops room we had just left,

"Captain, I have just been talking with the Met-man in Port Stanley, the latest METAR* is BAD. It's snowing heavily! There's low cloud with strong winds; it's a BLIZZARD..." I took a deep breath and pressed the transmit button.

"Roger...! That's all copied. Thanks a lot. OUT!" The co-pilot looked at me waiting for my reaction, and I said,

"You wouldn't do this job if you were superstitious would you?" we both laughed, then, ."..tell ATC we are starting engines. WE ARE GOING!"

Three hours later in the cruise, with just over five hours left to run, I managed to get through to Mount Pleasant Operations on our hand- held Iridium Satphone. It was always going to be a bit hit or miss, but from the few words I got back from a sleepy airman it seemed that it had finally stopped snowing and they were busy clearing the runway. I couldn't get any real sense of whether it was going to be done in time – he didn't seem sure himself.

At least I got the message across that we were coming! We passed 'the point of no return' heading south – it would now be a landing at Mount Pleasant or Uruguay. I said we should celebrate with a cup of tea and called the forward galley on the intercom; the stewardess was new. "Two teas please - Julie Andrews." Her puzzled response indicated she had not understood the Sound of Music reference, so I clarified for her... "WHITE NUN!" Then she got the joke and guffawed loudly down the phone.

It was no guffawing matter nearly 6 hours later though. We were well into a solid IMC descent towards the Falklands; "no fighter escort this morning." We flew through some really heavy snow showers and had the engine Thermal Anti-Ice system on most of the way down.

Occasionally the buildup of ice on the windscreen wipers and pillars outside the flightdeck indicated the wings were getting iced up too, so we hit the Wing Anti-Ice system – it was satisfying to think about all that ice being shed back there off those important leading edges. We broke cloud overhead the field and looked down on a very wintry landscape, there was whiteness as far as the eye could see... But there, down below us was one beautiful strip of black tarmac – the runway was clear!

As we shut down on The Pan*, which like the taxiways was covered in ice and snow, the relieved smiles on everybody's faces was worth all the effort and tension of the previous ten hours. Even the DAMO was cheerful

– amazing! We were acutely conscious that the weather 'window' would only be open for 12 hours at most.

After our minimum rest of another 10 hours, (including food and sleep in a real bed), we were back onboard our ship of dreams. There was no replacement crew down here and we needed to head north again as soon as possible – more snow was falling as we finished fuelling. I could see the overworked snowplows hard at it on the runway outside the flightdeck windows.

We completed our departure briefing and were all strapped in again, I looked across the cockpit at the FO smiled and said, "Mate...! As the Good Shepherd said... 'Let's get the Flock outta here!'"

* *TAF – Terminal Airfield Forecast*

* *METAR = METeorological Actual Report*

* *The Pan - military slang for movement/parking area for aircraft*

...One beautiful strip of tarmac - the runway was clear

24 THE JACKASS GENE

By Capt James McBride

A Disclaimer: *Nothing which follows in this story takes away from the fact that the vast majority of airline pilots are highly disciplined, committed, professionals who are dedicated to conducting their flights in accordance with the highest standards of safe operation. That being said, there are 'exceptions to every rule' and to those tasked with following them...*

When I think back, one of the first times I came across evidence of deliberate flaunting of aviation regulations was in the early 1980s during Military Flying training in the UK. It would be hard to imagine a more disciplined training environment for young aviators. Seriously, we were there to be pushed to the limits of our abilities and also to most of that of the aeroplane. Yet there was one gifted pilot for whom the challenge of just keeping pace with (and exceeding the requirements of), one of the most demanding flying training programmes ever invented was not enough.

Trainee Rob always needed more. Some of the things he did in-flight would be classed as incredibly stupid by the establishment if they had ever found out. He used to like to 'experiment'. So one day, while flying straight and level in the Jet Provost, he engaged the ground control locks which jammed all the flying surfaces in the neutral position. He then proceeded to 'fly' the aircraft just using the trimmers! Completely outside the design spec of the machine – the clue is in the name by the way, GROUND control locks.

Imagine the possible consequences if he had been unable to free the control locks and they had jammed in position? It would have been one thing to fly the aircraft on the trimmers in the open sky at medium altitude, but quite another thing to have tried to land the thing! Very few trainees would have taken the risk that's for sure, but Rob was 'special', always thinking outside the box. In fact he was the star of his course and passed the military training programme with honours.

The similarities between Rob's use of the control locks and the recent report I heard of the mischievous airborne application of the 'Parking Brake' in a B737 are remarkable. *"I wonder if it is possible to set it in the air...?"* The problem with the airliner however, was that once set; the pilots found that the Parking Brake would not release. Now call me old fashioned, but I come from the school which says, 'don't even use the parking brake on the runway waiting for takeoff clearance just in case it jams in the ON position,'

NOT TO BE USED IN FLIGHT!

I (and many others) always hold the aircraft on the toe brakes no matter how long the delay. Although very rare, there have been a few instances over the years when an aircraft has blocked the runway because the parking brake got stuck. Again, the clue is in the name, it is part of the aircraft braking system used for parking the aeroplane and nothing else.

A sensible car driver would not contemplate pulling on the handbrake while travelling at cruising speed on the motorway, so what possessed our heroes to do so in an airborne 737?

Needless to say it was no laughing matter and they quickly realised they were in trouble. They determined that IF they landed with the parking brake set, (mainwheels locked) they could burst the tyres in addition to other potential aircraft damage to landing gear and associated systems. So they took up the holding pattern over London Stansted and discussed the problem with the company's engineering department on the radio. After being talked through the process (involving pulling circuit breakers etc.), they managed to free off the brakes and land the aircraft without further incident.

Maybe the manufacturer should placard the lever *'NOT for inflight use'*

It would have been so good to have been a fly on the wall at the post-flight debriefing with the engineers who met the aircraft.
Although the above event was relatively recent, 'experimenting in- flight' with unapproved procedures has been the cause of unintended consequences for pilots through the years. There was a case I recall in a UK airline when the pilots of a B757 enroute from Manchester to the Canaries decided to experiment with pulling and resetting circuit breakers.

In the climb, soon after takeoff, the EICAS* screen came up with a crew alerting 'status' message R ENG SPEED CARD. The message (Right Engine Speedcard) was in Cyan on the EICAS screen and status messages would not usually have any detrimental effect on the progress of the flight. As they had already been dispatched, it would not prevent their scheduled arrival in Tenerife where the company had maintenance cover.

According to the captain, however, the message was irritating and they thought that perhaps they might clear it. So they agreed to pull out the R ENG SPEED CARD circuit breaker on the panel in the flightdeck to see if this had any effect" – nothing. So they put it back in again. As the c aptain said to me while relaying the story years later of what went wrong with this flight,

"You will have heard the words, 'Well... it seemed like a good idea at the time...'"

Disappointed that they had been unable to clear the message off the EICAS*, they then proceeded to pull the L ENG SPEED CARD circuit breaker, with similar effect, nothing happened and the message remained. They then had the bright idea of pulling BOTH circuit breakers together to see if that would have any effect – by now they were at cruising altitude, Flight Level 350 (35,000 feet).

What both pilots failed to realise was that the engine speed 'cards' have integrated computerised auto-functions with several other important systems. For example the speedcards monitor N3 turbine rotor speed inside the RB211 engines. If N3 falls below 45% RPM, the speedcard will detect/signal an engine failure. If both speedcard circuit breakers are pulled simultaneously, the system will indicate a dual engine failure...

Author showing the RAT - B757 on Flight Test Munich 2009.

This results in the immediate automatic deployment of the RAT. The B757 Ram Air Turbine is a powerful air driven motor, which drops down into the airstream from the aft right-hand fuselage fairing and produces hydraulic pressure for the Centre HYD system to power the flying controls. The EICAS system now indicated 'RAT deployed' much to the pilots' dismay.

There was an immediate reaction from the cabin too as the Senor Cabin Crew entered the flightdeck to say there was a,

"Huge noise and vibration now in the cabin under the floor at the back of the aircraft!" The passengers were complaining and it was disturbing the in-flight service. Restowing the RAT is a function which can only be performed by the engineers on the ground and yes, you guessed it, a diversion to London Gatwick was carried out to get this done.

Coincidentally, none of the stories of bizarre inflight behaviour which I have heard involved female pilots. In fact one or two of the best examples of gross stupidity were told to me by a female first officer with whom I operated some time ago. She was as horrified as I was by the irresponsible actions of our colleagues who were all male. I said to her,

FO's FMC with lifting handle raised – ready to lift out the unit.

"So you don't think any of the girls would be this daft then?" She was adamant in her reply,

"Oh my God! No way! We wouldn't even think of it." So I summed up by saying, Well I guess you girls just don't have the Jackass Gene then...!?"

She laughed at this and then told me the best story of all. This event just has to be the winner of Jackass of the Year prize and should be shared by both crew. The two pilots of a Boeing 737 (one of whom had a small screwdriver) partially removed the First Officer's FMC (Flight Management Computer) while inflight "to see how deep it went."

You can sort of see the attraction and why their curiosity could be aroused, the FMCs are both fitted with a neat little foldaway lifting handle on the top...

The problem was that they could not get it back in again! There are clips which need special tools to reset the device back in the hole it came out of. So they left it partially extended out of the lower instrument panel saying,

"Okay! Don't worry, we will restow it on the ground", which was all very well until they were on final approach and could not lower the landing gear... the protruding FMC unit was blocking the lever! They carried out a Go-Around and climbed away to the holding pattern, to try to get it back in. Finding this an impossible task, they eventually gave up and used the (emergency) manual gear extension system to lower the undercarriage to make a safe landing.

Apparently the engineering report made reference to 'witness marks on the top of the FMC, likely to have been made by the fire-axe...' (This part made me laugh out loud).

Then there was the story about the Low-Cost Airline pilots flying a Boeing 737-800. The year was 2018. Two captains, (one of them a line trainer) were tasked to ferry an empty aircraft to an airport in Poland. For some reason (possibly machismo related?) they decided to experiment with how short they could land the aeroplane. The first taxiway was a very short distance from the runway threshold. They both knew that with passengers onboard it was not normally possible to vacate the runway at that point.

So they set the aircraft up for a **very** 'shortfield' landing. They set the Autobrake system to 'Maximum' and briefed to use full reverse thrust on touchdown. As the airliner was empty, this was all unnecessary for landing safely at this particular destination. The two pilots had just invented the Bravado 1 arrival.

On touchdown and with the huge decelerative forces which their landing technique generated, a bar trolley from the rear galley ran forward, down the cabin and smashed into the flightdeck door. The noise in the cockpit must have been phenomenal. After shutting the engines down on the stand, the pilots realised they were unable to open the severely damaged door and had to call the engineers to come and let them out. The 737 was now unserviceable awaiting repair to the door.

They had neglected to ensure the galleys were secure preflight and after investigation, both pilots were summarily fired from the airline.

Moral of the story? Well if you are bored as a professional aviator while flying aeroplanes to the point of 'experimenting' in-flight – I suggest you consider another career. Alternatively in the words of our American cousins across the Atlantic, "If it ain't broke... don't fix it!"

Jackass was an American reality series, originally shown on MTV from 2000 to 2002, featuring people performing various dangerous, crude, self-injuring stunts and pranks. (Wikipedia Reference).

** EICAS – Engine Indicating and Crew Alerting System. An automated pilot information system in the B757.*

*

25 THE TURBULENCE SWITCH

By James McBride

As frequent flying passengers, how often have we heard those words on the PA from the Captain .".. and if you are remaining in your seat; please keep your seatbelt loosely fastened during the flight"? Of course we all buckle up for takeoff and landing and that is ensured by the Cabin Crew who conduct a 'Cabin Secure' check prior to these critical phases of flight, but how many passengers neglect to refasten their seatbelt after returning to their seat in flight? Quite a number I suggest.

Statistically, being hurt by the effects of turbulence in the cabin of an airliner is the single biggest cause of injury while flying, so perhaps more passengers should heed the wise words from the flightdeck. That being said of course, some types
of turbulence are difficult to predict. It is said that forecasting turbulence is more of an art than a science - Clear Air Turbulence (CAT) is especially difficult. As the name implies, this is turbulence which is invisible and not easy to anticipate.

It is often associated however with Jet Streams, especially near the outside edges of the Jet.

Imagine a fast-flowing river, the main core of which is quite smooth; however near the edges of the riverbank there will be all sorts of eddies and back-currents – this is where you will find the most turbulence. The core of a Jet Stream can commonly be moving as fast as 250mph – that's a lot of energy.

Large masses of air moving at different speeds can be forecast though and they are included in the computer flight plan (CFP).

The computer is programmed to look at the different wind vectors (direction and speed) above and below the planned operating flight level and by comparison with known data produces what is called a Shear Ratio. The Shear Ratio is normally printed on the CFP next to each waypoint as a single digit from 1 to 9, the higher the number the more likelihood of 'shear'/turbulence. In reality many pilots treat the information with some scepticism because after years of checking the Shear Ratios on hundreds of flightplans, I can say it is not a truly reliable source of information. Nothing exists in isolation though and if there are high SR numbers at waypoints which are close/adjacent to Jet Stream entry/exit points (as indicated by high altitude W/V charts), then they are worth believing.

In the cockpit, experiencing turbulence is no more enjoyable than in the cabin. It is categorised in the industry as light, moderate or severe. The best source of information regarding expected turbulence comes to us from other flights, especially those at our own cruising altitude. Often we will hear Air Traffic asking other aircraft to describe their flight conditions – these being called 'ride reports'.

I always smile when I hear the American crews calling back with ."..erm, we got a few ripples on the coffee here", or the very descriptive "Smooth Ride." Then there is "Yeah; Flight 123; we got some 'light chop' ...", this transmitted with all the nonchalance of a chap out in his fishing boat on a lake, although he is actually sitting in the pointed end of an aluminium tube travelling at 500mph, more than 5 miles above the surface of the earth.

One of the problems with descriptions of turbulence is that it is a subjective matter. Light turbulence experienced by one pilot may be interpreted as 'moderate' by another. This is particularly true with regard to new recruits who will reach for the 'Fasten Belts' switch perhaps sooner than more experienced operators. The older aviators know that the correct name for the switch is "The Turbulence Switch." Bizarrely and not uncommonly as soon as you switch it on, the turbulence goes away...

Of course after it is activated and the Fasten Belts signs illuminate, there is a reaction by the Cabin Crew who conduct their duties in the cabin to ensure that all passengers return to their seats, strap in and remain there. Not only that, but they will also suspend the cabin service, aswell as stowing trollies and service carts. All of which probably explains why some flightcrew are reluctant to switch on the seatbelts sign too early.

In reality, continuous light turbulence is not dangerous, although it can be unsettling for some cabin occupants. If experienced at cruising altitude, with light wind velocities present; nil adverse reports from other flights and no signs of Jet Streams etc., then it is often safe to say that a sudden encounter of moderate or severe turbulence is unlikely. Once the signs have been switched on; even if the bumpiness subsides and the flight becomes smooth again, it is a brave pilot who would switch them off again immediately.

In fact what many of us will do is start the stopwatch in the flightdeck and note the number of minutes since the last of the bumps was felt. We reckon that if it is still smooth after 5 minutes, then it is *probably* safe to switch the signs off again. If not, then 'the law of Sod' will prevail and in response to the belts being switched off, the turbulence will come back with a vengeance! Some days you get the feeling that you just can't win...

One of our difficult decisions, however, comes while cruising in high altitude cloud. Wispy/misty layers of high-altitude stratus which prevent us seeing what's coming ahead. Even with the weather radar activated and scanning our forward path, showing no significant returns, it is often wise to run with the passengers strapped in while there is any hint of turbulence. Experience shows that these conditions can produce some unexpected bouts of rocky road.

At other times though, perhaps it is aswell to switch on the belts sign as precautionary measure and if nothing else to reassure the passengers. From the customers' perspective, you can imagine them sitting in the cabin experiencing continuous light and very occasionally less-light chop saying to themselves (and each other), *."..don't those guys up the front REALISE it's bumpy?"* So, as a PR measure and as a precaution, occasionally it is worth activating the seatbelts sign even when from our point of view it is not strictly necessary from a flight safety aspect.

There is a growing school of thought these days, however, which advocates the use of the seatbelt signs virtually with any turbulence present in response to the modern trend in compensation seeking litigation. Immediately following the April 2017 Qantas Flight 23's encounter with turbulence when 15 people were injured, a lawyer claiming to represent some of the injured was making statements within hours of the flight landing safely.

The fact that most of the 'claimants' were not strapped into their seats on this flight into Hong Kong from Melbourne and that only one passenger was hospitalised, (later released) will be of little consolation to the company. The event is still under investigation by the Australian ATSB and we can be sure that the final cause analysis will be definitive. The dramatisation by the media reporting the incident will not make the investigators' job any easier that's for sure.

There was talk of, "passengers hurt in stick-shaker accident" which in itself mis-identifies the function of an essential safety system fitted to modern airliners to alert pilots of an imminent stall. The public could be forgiven for visualising the 'shaking-stick' of the control column as being a CAUSE of the turbulence experienced in the cabin – WRONG!

Also in some TV News channels' supposed recreations of the event; there was broadcast footage of a control yoke from a completely different type of aircraft, (assumed to be in a simulator) which was shaking and then magically pushed itself forward! The control wheel was from something like a Caravelle. It was the inverted 'W' type (think Concorde or HS125) and was obviously reacting to a stick-pusher system NOT fitted to the subject B747. *Dear God, where do they get these so called 'experts' from...?*

There were additional reports from passengers that the captain referred to "Wake Turbulence from another aircraft", during the PA announcement which was made after the event. This is a possible cause of sudden turbulence to an airliner inflight; however it would have to be another 'heavy' type (similar in size to a 747) and not far ahead to have such an effect on the flightpath of QF23.

Usually ATC ensures that preceding aircraft are at least 5 nautical miles away from traffic behind at the same level. As with all aviation incidents, it is wise to wait for the publication of the complete findings of the investigative team rather than speculating as to the cause. However be assured, the shaking in the cabin will not have been caused by any 'stick-shaker' mechanism!

We live in modern times and the flight-planning/weather-forecasting systems used in airlines have never been better. In addition we have advanced Weather Radar and TCAS* systems which help us identify where other aircraft are flying in the skies. Used carefully they can also indicate to pilots safe areas to operate ahead when you can see flights which are maintaining their tracks even though weather returns are close. Even with all the technological advancements, I'm betting that 20 years from now... it will still be referred to as "The Turbulence Switch."

** TCAS – Traffic Collision Avoidance System.*

26 SWISS CHEESE DAY

By Capt James McBride

Italian ATC Tower

"ALITALIA XXX! WHICH SID are you flying?" The voice of the radar controller on the VHF sounded more than a little concerned, he was mighty pissed off, you could tell.

In our own flightdeck we looked across at each other with raised eyebrows – this sounded serious. We had taken off from runway 26 at Catania and were climbing through Flight Level 90 on the PEKOD 5B route, the Alitalia Airbus behind us should have been following our track...

The reply we then heard from the A320 was,

"We are flying the PEKOD 5 Bravo departure, Alitalia XXX..." I noted that the tone of his transmission was confident, yes he really did believe his aircraft was flying that SID, (Standard Instrument Departure) but the Air Trafficker was having none of it. Clearly he was watching the radar display and could not believe what he was seeing.

"NEGATIVE ALITALIA XXX, you are 2 miles NORTH of the airport and have turned 180 degrees towards the East – *this is NOT like any of our standard routings!."*

The Italian Captain then replied somewhat hesitantly, "Roger Sir... but we think we are on the PEKOD 5 Bravo SID... We shall file a report Sir..." to which the controller shouted,

"I WILL FILE A REPORT ALITALIA, turn right radar heading 110 degrees and CLIMB Flight Level 240!"

There was then more discussion on the R/T, but we missed the rest of it as we were handed off to Rome Control on the next frequency in our climb to the North. Bear in mind that SIDs* are designed to keep IFR departing aircraft separate from both terrain and other air traffic. Even a slight track deviation from the published SID is a potentially serious matter, but to completely diverge from the intended track is almost unheard of.

We were puzzled as to how it could have happened. You see the East/West runway at Catania airport in Sicily is located in a rather special place. Just 15 nautical miles north of the airport there is high ground rising steeply to over 10,000 feet above sea level – the actual Minimum Safe Altitude (MSA) is 13,300 feet. Mount Etna has the distinction of being listed as the tallest, active volcano in Europe. Certainly when 'considering the threats' during preflight briefings from Catania, Mount Etna always gets a mention.

Mt Etna - the tallest, most active volcano in Europe.

All the more worrying then, that the errant Alitalia flight turned to the north, towards the highest terrain closest to the runway, but why would they do that? Clearly it was not the fault of ATC (according to what we heard on the radio) and these days most unlikely to have been a technical malfunction, so it was most likely to have been a Human Factors error. Once we were up at Flight Level 380 in the cruise we had time to discuss the event properly.

We ran through the sequence of events leading up to our departure, taking off only two minutes ahead of the Alitalia. I recalled that it had started as one of those classic Swiss Cheese Days. When you realise there are a multitude of possible 'Gotchas', waiting for you as you report for duty on a busy 4 sector day. In the crewroom, our preflight briefing was interrupted when the first officer informed me that the fourth of our flightplans was missing from the computer system so valuable minutes were taken up chasing it with the remote Flight planning Department.

Almost simultaneously the cabin crew informed me that our Senior Cabin Crew Member (the Number One) had not reported for duty yet, but had phoned from the carpark saying she was waiting for the shuttle bus. She would arrive a few minutes after official report time. In the interests of good crew relations, (modern CRM) I try not to overreact when I hear things like this, although it is an irritation. Finally we got the briefing done; the crew altogether; legal documentation all present and headed out for the jet, only 5 or 6 minutes behind the drag curve. We were technically running late.

The crewbus which had been ready to take us to the remotely parked aircraft, if we had been ontime, had just left to do another job so we have a further wait for another bus... While we stood idle, I talked with the first officer and we discussed the fact that Mount Etna has been more active recently. These days we aviators are very conscious of the danger of volcanic ash clouds and it was only in December 2015 when Mount Etna last erupted with consequent disruption for air traffic in and out of Catania airport.

Of course there are contingency plans to permit flight operations to continue in the case of major volcanic activity, but it does add another layer of complication for the flying programme. We arrived at the aircraft to find that the previous crew had sealed it all up, so it took a little more time than normal to unlock it all and get the AC electrics established. Thankfully the refueller was there ready for us, so I took the opportunity to get the dispatcher to call for the passengers to board the buses.

While the FO started preparing the cockpit for his departure (he is Pilot Flying for this sector) I carried out the walkround and ensured the refuelling was happening. Now the cabin crew were well into prepping the aircraft and called my attention to look at a defect in the cabin,

"Captain, can you look at this please, is this okay for our flight?" One of the aft lavatories would not flush and it is not in the aircraft Techlog... What to do?

"Yes, no problem, lock it off please, we are not full of passengers and it is only a short flight." I reason that there are two other lavs also, but it all takes time...

Why is a toilet flush problem important to us? Well you only have to look back at the Air Canada DC9 accident (Cincinnati 1983) when a malfunctioning toilet flush motor started an inflight fire which resulted in terrible loss of life. In our case if nobody can use the lav, the problem is solved. As an extra precaution I ask the engineers to disable the lav and we can get it fixed later – it is an acceptable deferred defect.

Some smart cabin crew actually ask the pilots to keep the cabin cool in winter as the passengers come aboard – then they will sit down quicker as they keep their big coats on.

Fuelling is now complete and passengers are boarding. The dispatcher is looking at his watch and I am also conscious of the Alitalia Airbus on an adjacent stand. They are boarding also, so they might delay our pushback. I ask the cabin crew to expedite the boarding process as much as they can, but I am aware there is a limit to how fast you can get people to sit down with their baggage stowed away.

Some smart cabin crew actually ask the pilots to keep the cabin cool in winter as the passengers come aboard – then they will sit down quicker as they keep their big coats on, rather than obstruct the single aisle while taking off their coats, scarves and hats etc. Another problem occurs now when the Ground Power Unit disconnects without warning and the aircraft lighting swaps to the standby electrical system. In the flightdeck we switch on the APU, but it does take a minute or two to come online... "Yes it's one of those days" I say to myself.

Now we are focussed on the final preparations for flight, the fuelling is complete, the Techlog is signed off, and we are running through the preflight checklist. We discuss the threats and are conscious that there is a tailwind for takeoff – we are planned for the PEKOD **5A** departure for runway 08. As the old sailing masters used to say, "The win... Ah, she is a fickle mistress, " Air Traffic Control are still using runway 08 for departure with a 6 or 7 knot tailwind from the west. We know from experience that this is more convenient for ATC traffic flow and is the preferential runway.

In fact it is relatively safer in some ways, even with a tailwind, because the takeoff run is straight towards the sea - the Departure End (DER) of Runway 08 is nearly on the beach - so there is no problem with terrain obstruction in the event of an engine failure on takeoff. The last few moments of our leaving from the terminal come in a real rush. The pushback team arrive with the tug and hook up the towbar to the nosegear, the Senior Cabin Crew confirms all pax onboard, the dispatcher rushes out with his signed copy of the loadsheet, the cargo holds are closed and the front steps are retracted. I do a quick PA to the cabin, then the FO obtains clearance for push and start and we complete the last of the prestart checklist...

Phew! Only a few minutes late and we are pushing back BEFORE the neighbouring Alitalia very pleasing. As we are starting the engines however, ATC is trying to give us our departure clearance on the radio, but I ask the FO to reply,

"Standby," we must keep our priorities in order. It would be a shame to 'cook a motor'* while we were distracted copying down the SID, stop altitude, transponder code, airborne frequency etc.

Similarly, these are important items to ensure that we get right, so it is vital that BOTH of us listen and agree with the ATC clearance and are not distracted by simultaneously monitoring the engine start process. Finished with engine starting and now we are stationary, waiting for the groundcrew to disconnect all from down below,

"Go ahead with the clearance" is the transmission from the right-hand seat and ATC reveals that they have changed runways! Now... it will be RUNWAY 26 for takeoff and the PEKOD **5B** SID. We spend a couple of minutes making all the required reprogramming changes in the FMC, plus resetting all the navaids then we call for taxi. Our company is quite prescriptive about what to do in the event of a late notice runway change and for this we should be grateful.

Thinking about it later on, that was probably where the Alitalia guys got it wrong. Pushing back only a couple of minutes after us, they will have been faced with a similar situation, but we would bet money on it they forgot to change the SID routing in the flight computer. As soon as they were airborne and the autopilot was engaged, it simply turned the shortest way towards the PEKOD 5A magenta track – with a wind from the north, the aircraft heading would have been slightly northwest after airborne.

** 'Cook a motor' – slang term meaning to allow an engine to overheat during start-up.*

*

27 GAME CHANGER

By Capt James McBride

As an airline pilot there comes a time when you upgrade from First Officer to Captain. It is an incredibly stressful period of your life when, not only must you be able to fly the aircraft well, that's a given, but also you become Manager of the Team. When your Command Course has been successfully completed and you are "cleared to the line", overnight YOU are the one the team turns to when things go wrong. Of course, 98% of the time they don't. Everybody comes to work on time and does their job well. The aircraft operates exactly as the manufacturer intended. All the passengers arrive in good time for their flights and behave in a normal, civilised manner.

In addition, over 90% of the time, the weather and ATC all permit unhindered, on-time airline operations. In fact I have a theory that we are not really paid for these times. We actually earn our salary on the rare occasions when things go wrong. When engineering malfunctions rear their ugly head; when passengers go missing; when ATC make mistakes; when passengers become troublesome or disruptive; these are the times when an Airline Commander earns his/her money.

In the majority of flights, we go to briefing, ensure that the crew is all present and correct, (well prepared for their duties) perform the preflight checks, order the fuel and board the passengers. In a well-structured airline company everything else takes care of itself. The support crew do their jobs in the right place at the right time, ATC clears us to push and start right on cue and we (the operating crew) get to do the fun part of our job, which is to fly the jet.

I know that all sounds simplistic, but honestly, most days that's just what happens. This is exactly why Airline Travel as a form of transportation is incredibly safe. Statistically the safest form of mechanised travel on the planet. The airline system depends upon all personnel playing their part in accordance with the rules of the game. This produces an on-time operation which is executed to the highest standards of safety.

When the routine is repeated so often, pilots can be forgiven for becoming a little blasé about their profession. It is important however to be able to recognise at an early stage when suddenly the rules of the game have changed. Those are the times when quite possibly the Aircraft Commander is required to earn their keep. Sometimes you can anticipate a game-changer. For example, on arrival at the jet, you check the aircraft Techlog and find that the machine has been downgraded such that it cannot be permitted to fly an Automatic Landing from a Low Visibility approach. You are already aware that the destination airport is operating in fog and low cloud so this function is essential... The game has changed.

Now you need to coordinate with the Ops Dept and make some changes to the plan. Of course this is a simple one and in most cases Ops should have already spotted this 'showstopper' and will have swapped the aircraft. But if the downgrade occurred during the last flight, they may be unaware. There are many similar combinations of hardware and environmental factors which impinge upon the smooth (and timely) operation of airliners.

Occasionally they make big news and the airline must alter commercial strategy to overcome the barriers. In our jet on the line however we are not so much concerned with the Master Plan as seen by those in HQ. Our challenges are usually of a tactical nature. The longer you do the job, the more you are aware that you haven't seen it all. This is especially true when it comes to software issues - dealing with people. Of course usually the pilots don't get involved in passenger problems at all. For sure when the doors are closed and we are all sealed in our pressurised aluminium tube, we never see passengers at all these days. If there is a problem with one or more disruptive passengers in the cabin, then the flightdeck door remains locked.

Before flight, while we are all on the ground it is a different matter however. Very recently on one of my flights I was in the flightdeck with my First Officer when the Cabin Crew came in to give us a game-changer. We had not yet started boarding the pax. The passengers had been brought to the doors of the aircraft, front and back to queue while the crew completed their preflight security checks. The high-vis security straps were fitted across the entrances indicating that nobody was allowed to board. One of them had different ideas however and ducked under the strap at the back of the aircraft. He walked into the rear galley. He was a well-dressed gentlemen aged around 40.

When the young steward challenged him and said he must vacate the aircraft again and wait outside, the man was rude (verbally abusive) to him and even "shoved" him bodily, trying to push him out of the way. This was related to me by the Senior of the Cabin and had been witnessed by another crew member. After the Steward finally convinced him to vacate the aircraft, informing him that he would not be allowed to travel, the passenger then stood at the rear door of the airliner preventing the crew from boarding other passengers. He was not in the least bit repentant.

"Oh dear... A game-changer!" I said and immediately left the cockpit.

I made sure that the FO continued to prepare the aircraft for flight and then stood with the Senior in the forward galley. I got her to repeat exactly what had happened and then assured her that this passenger would not be flying with us today.

"Leave it to me, hold the boarding until we have got him removed by the police." I was confident that provided he did not have a hold bag, the delay would be only a matter of minutes. I was actually very pleased that this disruptive passenger had chosen to misbehave on the ground. I had an easy solution here.

I arrived in the rear galley and checked with the steward who had been pushed,

"Point him out to me please?" The passenger was standing just outside the rear entry door at the top of the mobile steps. I had no intention of getting into any discussion, that would just waste time. We only had 20 minutes before scheduled off blocks. I went straight to him in my uniform and asked him for his boarding card and passport making sure that I smiled as I asked for them.

It's a funny thing, but a smile is the most disarming thing you can do in these sorts of situations and shows you have confidence. I also was smiling because I knew what was going to happen to him, but he didn't. He handed over the documents promptly and I said to him quite clearly and politely,

"Please wait there a few moments Sir, this won't take long..." And then I disappeared back inside the airliner. I walked quickly up the aisle to the front of the empty jet and looking out of the forward entry door I could see the Ramp Agents down below on the concrete. I went down to them and said in a clear slow voice,

"GET THE POLICE, this gentleman, here are his documents is NOT PERMITTED to travel today! As soon as he is removed from the aircraft you can commence the boarding, so let's get on with it!" Again I gave the Dispatcher a big grin and turning to climb the forward steps, I could see her bringing the Gendarme. Looking backwards to the top of the rear steps, I could see the offending gentleman looking at me and he had obviously seen me hand the passport and boarding card to the Dispatcher. I gave him a cheery wave and then pointed at the approaching Police Officer.

Back in the flightdeck we nearly managed to catch up the time lost and in fact we were only 6 minutes behind schedule as we pushed back. The first officer was surprised that it had all happened so quickly, but I emphasised that we had to act fast to keep the flight on-time if we could. Also as I said to him, part of our job is educating the travelling public where it is necessary to do this. For example the offloaded passenger would never be rude to a crew member again and nor would any of the pax who had witnessed the event.

For us as a crew to have acted differently and tolerated such abuse sends the wrong message. Finally, when our cabin crew look to their leaders (the cockpit crew) for support, we should give that support whole-heartedly where possible to do so. These days that is mainly on the ground.

The important thing for the Captain to be sensitive to, is to know when the game has changed. In this case the scheduled time of departure, (STD) which is usually the most important part of our focus up at the front of the aircraft was suddenly secondary to resolving an event which had occurred right at the back of the airliner.

28 TO TAXI OR NOT TO TAXY?

By Capt James McBride

It is an activity which I have been involved in for many years, but I was never sure how to spell it. The movement of an aircraft on the ground (or close to it in the case of a rotary wing machine) under its own power is called to "Taxi." It is a verb, a doing word, however there is some dispute about whether the word "Taxiing" exists at all. There is an alternative spelling used by many people and that is "Taxy" – same meaning. Whatever it is an important part of operating aeroplanes, especially large transport aircraft.

In fact, it is a critical part of flight whether taxiing in after flight or taxiing out before it. It may seem like a statement of the obvious, but aircraft are designed for flight. A flying machine 'on the ground' is really out of its natural habitat. All those streamlined edges, smooth lines and large extended flat bits called wings are very vulnerable to damage while not flying.

One of the biggest problems with airliners for example, is that they must be serviced by groundcrew with machines. These machines by necessity must come very close to the fuselage, wings and landing gear. As the machine approaches the airframe all manner of things can happen. Maybe the driver's foot slips off the clutch and the belt-loader (for baggage) surges forward and rams the hull creating a huge dent or worse. Possibly the pushback tractor has a malfunction while being attached to the aircraft and causes injury to the nose-gear. Even in good weather attaching towbars/tugs to airliners is fraught with danger to both man and machine; imagine how it is in bad weather, middle of the night with a tired/exhausted crew.

Even in the hangar, where you could expect the aeroplane to be safest, deterioration to the airframe can occur. Sometimes this is caused by other aircraft being manoeuvered in close proximity. The term often heard to describe accidental mutilation within the haven of the plane garage, is "hangar rash." This expression implies that it is some sort of medical condition which is potentially infectious and hard to prevent.

Accepting that there are many ways in which aircraft can suffer harm while stationary, brings us to what happens when they start to move under their own power. That's when the real fun begins! To taxi an airliner from the cockpit has been described as 'trying to steer a semi- detached house while looking out of the bathroom window'. You are unable to see the rest of the house – apart from the window ledge, but you are aware that there is a huge amount of structure behind you as you follow the painted yellow taxi-line.

From the flightdecks of most airliners, it is not possible to see the wings and certainly not the tail. Technology comes to our aid however with cameras which can show various angles from the airframe, but there is still no substitute for the mark one eyeball. You still have to lookout. It is funny, but in a machine which is built to travel across the earth's surface at 500mph, it is easier to get lost at 20mph while at the airport.

Part of the problem is that it is the bathroom window effect - you have such a narrow view of the world, to perform what is quite a complicated operation. Also, your view is horizontal, but the plan of the airport and taxiways is from above. With many airports the taxiway markings and signage is clear, although some are not so good. It is a trap for the unwary when you expect to see signs in the grass alongside the taxiways which are not there. In some of the more remote locations there are hardly any indications of which turning to take.

In Thessaloniki for example (Greece's second city) they appear to have run out of yellow paint! Taxiways, entries and exits to/from the apron are unmarked etc. etc. The hilarious thing is that Air Traffic Control still refer to "...take Charlie entry point to the parking" but there are no markings to indicate where Charlie is Even at international airports like New York JFK it is easy to take a wrong turning. Like we did with the Alitalia 767 after landing from Rome in 1996... I was the captain, and this happened on my watch, but in the right-hand seat was a VERY experienced ex-captain who was flying as the First Officer. Bear in mind that one of the critical functions of the non-handling pilot while moving on the ground is to guide, monitor and advise the Handling Pilot while they taxi the aircraft.

They normally do this by paying close attention to the taxi-chart and tracking the aircraft's position. Literally the HP will have his/her hands full of maybe 300 tonnes of live machinery which they are trying their best to move smoothly and gently with the tiller which is linked directly to the nosegear steering. On the B767 there was only one steering tiller which was on the captain's side, so the RHS pilot was always in the monitoring role. On this day, we landed and taxied off the runway with a long routing to our parking stand. Jack, my right-seater (not his real name) was in relaxed mood and so was I. After all, we had just landed in the 'Big Apple' after our non-stop flight from Rome and were looking forward to a couple of days relaxing in the city that never sleeps.

Well, we acknowledged the clearance and started taxiing, but my very experienced chap to the right of me, was chatting away about previous New York trips he'd been on. To be honest, both of us let our guard drop. There is a well-known phenomenon among pilots, that "the brain dumps everything after landing", but it is a time when you need to be still vigilant. Remember it's a critical phase of flight.

Anyway, to cut a long story short, we took a wrong turning on a taxiway with our 180-tonne wide-bodied jet and found ourselves travelling parallel to the one we needed – YIKES! We (with me steering) had taken a 90-degree right turn, the one before we were supposed to take and now we were heading up a cul-de-sac. When we both realised our mistake it was like a cold shock and I cursed myself for letting it happen. Being JFK, the ATC ground frequency was full of chatter. We knew there would be a problem getting our transmission understood, especially as it was going to be difficult to explain why a scheduled flight from a National Carrier had made such a schoolboy error. We were still moving at 20 knots or so when up ahead I saw there was an intersection with a crossing taxiway, the crossing was huge...

"Here Jack! We can turn it here and do a 180..."

I slowed the ship right down to walking pace and we both scanned it carefully, yes there would be room here. It was not a recognised turning place, but it would do. I steered the left maingear as close to the edge of the left side as I dared then with some asymmetric engine thrust, I wound on full right lock with the tiller. Those 767-300ERs could certainly turn tightly when you needed them to. In retrospect we had lots of room, and within less than a minute we headed southbound on the same taxiway. Any second we expected the call from a sharp-eyed Air Trafficker, "HEY! Alitalia 767 where ya going? Didn't you make a wrong turn just there...?" But nothing... literally nobody noticed. It was like our massive airliner was invisible as we popped back out on to the correct route and a few minutes later shut down on the stand.

It was a big lesson to learn however and I know that some of my colleagues look at me a bit weirdly these days when I say to them quite assertively "FOLLOW THE TAXI CHART, PLEASE...!" As we make our way through unfamiliar airports and even familiar ones come to think of it.

So.... is it Taxi or Taxy? After 35 years I still don't know how to spell it – does it matter? Well I guess not. What really matters is that we don't take for granted that we know where we are or where we are going. I will often say to my fellow cockpit occupant, "When it comes to taxying, don't trust me; because I don't trust me!"

At JFK - ATC Ground frequency so full of chatter – it was like our massive airliner was invisible... literally nobody noticed.

29 GEAR DOWN FERRY FLIGHT

By Capt James McBride

I distinctly recall thinking, *'there are downsides to promotion in the airlines...'* You see, when you start flying airliners for a living as a First Officer, your first major promotion is to change seat and get your Command. Some airlines refer to this as an upgrade to captain, whatever; it is the same thing. After flying with the company for some time as a captain the next step up is to become qualified as a Line Training Captain (LTC) and after that your aim maybe to become a Simulator Trainer and then Examiner (TRI and TRE). At some stage throughout all of this process you may in fact become a management pilot.

Different companies have varying views on the pilots selected for management positions. Traditionally it was normal for managers to be selected from the ranks of experienced trainers/examiners. This is still the case with most airlines because choosing from your experienced trainers is a wise move - they automatically have a certain amount of status and respect from their work with the pilots flying the Line. The downside I was thinking about, was that because you are manager/ trainer the company expects a lot more from you, certainly in the decision-making department...

I was facing a unique problem and I knew it. My Chief Pilot had called me early on a Sunday morning at home (my Day Off of course) and tasked me to go and ferry an empty aircraft to maintenance. It was a B737-300 with EFIS, the type was the mainstay of our airline and I was familiar with it. Not only that, but I was a Trainer and management pilot for the company – so you could say the airline put its trust in me to make the right decisions in the interests of the operation. I should have guessed this was no ordinary mission though when he said on the 'phone,

"Get it done, as soon as possible James, we need that airframe back on the line today. We have cancelled several flights because of its being out of service." I knew our young company was running a tight flying programme; we needed every ship that would fly.

When I arrived at the airport, I realised that this machine was the aircraft which had diverted a day previously with a landing gear problem. The issue had been that they (the pilots) were unable to get 'Three Greens' for landing at our company's base airport 15 miles away, so they diverted to Manchester international airport nearby which had a huge runway. This was a sensible move because as they had 2 Greens and 1 Red for the gear indication there was a possibility of a partial gear collapse on touchdown.

The crew spent almost an hour still airborne trying to solve their problem – they tried everything to get 'Three Greens'. Having carried out all the procedures and drills however, they finally came in to land still with one red light, an 'unsafe' configuration - the starboard main gear was theoretically not down and locked... Everybody held their breath as the airliner made its final approach, the fire trucks were all manned up, waiting at the side of the runway; ambulances and paramedics were standing by, blue flashing lights everywhere.

Luck was on their side though because on touchdown, the gear light on the right-hand side suddenly changed from Red to GREEN! A great result. This was especially so for the passengers. They had all been briefed for a possible main gear collapse on landing and were all in the *"Brace for Impact"* position. If necessary and as a precaution, the crew had prepared the cabin for a possible crash-landing and even a subsequent passenger evacuation. Fortunately as the landing had been 'normal', the passengers were disembarked as usual on the gate and the aircraft handed over to the company engineers. Now came the issue... the engineers couldn't find anything wrong with it. The crew was debriefed and both pilots were adamant that there had only been 2 greens prior to making the landing. On arrival the next day, I spoke at length to Gary one of the senior engineers with whom I had worked before.

He stated that the best option from the maintenance point of view was to get the aircraft back to our home-base and then they could jack it up in the hangar and carry out 'gear retraction tests'. He was also quite certain of what type of flight was required,

"It'll need to be a ferry flight with the gear down James, as we don't know what's up with it. You don't want to retract the undercarriage and risk another potential gear collapse landing..." So I called the Operations department and spoke to the Ops Controller there. I explained the problem and asked him for specific runway take- off performance and a flight-plan for a gear-down ferry flight with an empty 737. That was when I realised we were going to have a problem as the controller responded with,

"Sorry Captain, we can't do that we don't have the performance data. We can provide you with a normal flightplan, but not one for a flight with the landing gear down..." I was surprised, and said so, but he was adamant, ."..the computer won't run a CFP for a gear down flight, we would have to get authorisation and specific performance from Boeing in Seattle."

I looked at my watch – it was 1130am on a Sunday morning in the UK. *Hmmm... 0330am local time in Seattle on a Sunday morning. There's no way we will be getting any response from them till tomorrow... what to do?* I went back to speak with Gary and told him the bad news, his face fell. He said,

"She needs retraction tests on the jacks in a hangar, we think it's a rigging problem, but we're not sure. One thing is certain though; if you bring the gear up, they might not all come down again as it is right now..." I asked about the possibility of getting the aircraft into the hangar at Manchester and the response was negative. No company was prepared to help us without great expense and even then if we put it in somebody else's hangar, we would be at their mercy – that's the way of aviation. We needed to fly it out.

About this time, the First Officer arrived whom I knew well as an experienced 737 pilot, we worked together at our base only 10 minutes flying time away. We discussed the problem together and analysed all the different scenarios. One thing we were in total agreement upon was that we did not want to be the crew who landed the plane back here with the gear up or partially deployed and blocked the runway at an international airport for half a day while they craned the aircraft off.

The flight would be at least 15,000 kgs lighter than normal as we had nil passengers/payload and only a light fuel load. I recall saying,

"Look on the bright side, what's the worst that can happen? We have an engine failure on takeoff, but because the jet's so light, we will have loads of surplus power to climb away even with landing gear extended. And what are the odds on us having an EFATO on the very day we decide to fly with the gear down for 10 minutes...? It's gotta be Nil right...?" He nodded and replied,

"If we take some extra gas for the increased fuel burn due to the drag, then that would be okay."

I then had another call from the Chief Pilot on the company mobile and excused myself to go and take it. He listened carefully to my summary of the situation, including the problem with getting a flightplan and then said,

"Well James you must do what you think is best..." I shook my head as I closed the 'phone and muttered,

"Just like a chocolate teapot... no use at all!" and saw Gary the engineer looking at me, he said,

"You won't want to hear this either James. I've just come off the 'phone to the Chief Engineer and he said it's gotta be a gear-down ferry flight, preferably with the landing gear locking pins wire-locked in place!" Well I could see the sense in that. From the maintenance point of view they wanted to take no chances that the gear may collapse – it would be like flying a 50-tonne tea trolley, quite a novelty!

The FO was in the flightdeck preparing the aircraft for flight and I came in and closed the door, we had a long discussion and we were in agreement. We knew the flight could be safely executed and decided on the best course of action in the circumstances. I called Ops and asked them to send us a flightplan for the ferry flight, the controller once again emphasising that it would be for a normal CFP for an empty aircraft with normal config. – i.e. gear up. I said, "that will be fine thank you, file it in the system with a scheduled off blocks time in twenty minutes from now." I turned to the engineer and said to him,

"Tell the hangar we are on the way, they will have their aircraft in about 35 minutes time." He smiled and handed me back the Techlog for my signature. We secured the cabin for the ferry flight and closed the doors.

Our detailed briefing for the flight, included arming both forward doors as usual. As one very senior pilot told me once, ."..that way, if you're in a hurry exiting the flightdeck, whichever way you turn you will get a slide!." We covered all the possible threats and verbalised the management/mitigation of them, I switched off the mobile 'phone and asked him to call ATC for start-up clearance.

A few minutes later we were thundering down the runway at the start of a very short sector,

"Vee One... Rotate!" he called as I applied back pressure to the control column and felt the nose rise in response. *She's so light, she's climbing like a homesick angel...* I thought as we turned towards homebase, this was despite the fact we used the absolute minimum thrust possible for the takeoff. A few short minutes later and with all the checks complete, we were on finals and cleared to land at home base – *reassuring to see three greens this time...* I mused.

The landing was uneventful and we taxied in to park right outside the maintenance hangar. The doors were wide open, the big hydraulic jacks in place and there was the tug and towbar waiting for us. Even better was to see the happy smiles on the faces of all the engineers.

All checks complete, on the flightdeck I smiled at my colleague and we shook hands, "That was fun!" I said and he agreed.

I completed the Techlog, '*Empty Ferry Flight completed to MX base'.* Did we retract the landing gear?

Nobody will ever know.

*

B737 Maingear downlock pin with flag – "remove before flight..." Usually.

30 COULD A PRIVATE PILOT LAND AN AIRLINER?

By Philip Johnson, PPL.

I've always been interested in aviation. At 13, I learned to fly R/C model aircraft, which I flew until I was 26. In 1998 the job I was leaving, bought me a trial lesson at Westair at Blackpool Airport as a leaving present.

About a year later I decided I was going to learn to fly. I did 18 hours at Blackpool, which wasn't that good because I was there all day and they didn't give me the flights. The instructor even said,

"You aren't capable of doing more than two, half hour flights per day" when I complained that I was sat there doing nothing all day. The remainder was at Jersey Aero Club in Feb 2000, after 3.5 weeks coming back with all the remaining exams, 55 hours, and a licence.

Philip Johnson PPL at the controls - could he land an airliner?

I like the idea of simulators on PC, but I just can't get on with them, memorising keypresses just to turn my head so I can look where the runway is for example, takes too much time. I haven't flown anything bigger than a Mooney or Cessna 172 on the sim, although I did fly the night before my qualifying cross country – Jersey – Vire – Renne – Dinard, Jersey. At Dinard the French fireman at the bottom of the control tower said in broken English,

"That landing was rubbish." I replied,

"You didn't even see it, you were sat in your fire engine!"

By the year 2008 I had just under 300 hours total time and had flown C150, C152, C172, PA28-140, PA28-151, PA28-161, TB9, TB10, TB20, most of the hours on the TB10 G-BLYE. I had my night and IMC rating (including approximately 20 hours IMC).

I was married to Tracy who didn't really like flying, but would come with me provided she could shut her eyes! She is the same on a jet but it doesn't stop her from flying.

Personality wise, I'm fairly average, I'm generally happy in my own skin, am halfway between introvert and extrovert. On the scale of calm to neurotic, I am off the scale calm. You can't do anything to offend me, but if you try really hard I might eventually work out we can't be friends or workmates and I'll withdraw. I'm conflict averse and very easy going.

The Challenge.

Around 2009, I read an article in Pilot magazine from retired airline pilot Bob Grimstead. In this piece he stated that if a Private Pilot (PPL) was on the flight deck of an airliner at 35,000 feet, they would accidentally click off the autopilot while trying to call for help if the flightcrew were incapacitated. Then he said that with the aircraft being so sensitive up near 'coffin corner', they would be overcontrol the plane and make a big hole in the ground directly underneath.

I considered this was an insult to private pilots and I sincerely believed he was incorrect. PPL's do not usually fly automated aircraft, we don't fly with CAPS, Synthetic Vision, Garmin Autoland (the latter two weren't even around at that time). We are more likely to fly an old Cessna 150 that when it grows up wants to be a helicopter (i.e. it won't fly in a straight line at any power setting or airspeed, certainly not hands off). So I thought as a rule, PPL holders are good stick and rudder pilots, once they have their licence they are let loose in the big wide world and it is up to them to manage their currency and attitude. With no SOPs to help guide them, they are largely on their own, but that doesn't change the point – which is I have never met one who is a bad stick and rudder pilot.

So I wrote to Bob Grimstead and over the course of a few emails I put it to him that, "the average private pilot would reach an airport and within reach of the emergency services, somebody would survive," his reply was "If you are so bloody cocky then, prove it."

Some time passed, but I had not forgotten our email discussion. Move on to 2013. My wife found I think it was called Real Simulations at Harrogate. They had a full-motion Boeing 737 an F4 fighter jet and a Formula 1 car. She booked me for an hour in the Boeing 737 so I could lay this to rest once and for all.

"By the way, is there anyone on board who knows how to fly a plane?"
From the movie "AIRPLANE!" 1980.

The Big Day.

I went to Harrogate and met a Boeing 737 captain who had lost his medical. I explained to him my situation. I wanted to see what I could do on my own with no tuition or advice. He was adamant that I did not need to feel the controls at 35,000 feet, he said.

"It is sensitive, but you would fly it. I want you to start at 20,000 feet so that we can get you down in half an hour, the remaining half hour I'll show you some things and how well this simulator flies. And don't worry about knocking the autopilot off by mistake, it's over there." He pointed to the switches on the panel in front of the first officer's seat. He continued with,

"The transmit switch is UP to talk outside, and DOWN to talk inside so all you are likely to do is give a mayday call to the cabin crew." I said,

"Whoa! Please don't tell me anything else, I'm trying to prove something here." He turned the autopilot off and started the sim. Ok here goes...

"Mayday, Mayday, Mayday, this is G-ABCD a 737 with a private pilot in command, request help and radar vectors to an airport, position unknown, Flight Level 200, heading 180."

I got what must have been, given the circumstances, the shortest reply in the history of aviation, "Golf Charlie Delta, roger Mayday."

Then a bit later...

"Golf Charlie Delta, turn left heading xxx."

I set the heading bug, and turned with 30 degree angle of bank onto that heading. *Wow, this is nice, I don't have to do anything with my feet either...*

"Golf Charlie Delta, turn left heading xxx."

Another one of these and he has made me fly in a triangle.

"Golf Charlie Delta you are identified, your position is 72 miles north of

Aberdeen, weather at Aberdeen is CAVOK, suggest you head for there."

"Wilco Golf Charlie Delta."

There followed a number of instructions, each time I set the heading bug and turned onto the heading.

The first instruction started with a descent, into cloud. There was what I think you would call 'light chop' which continued while in cloud and for the rest of the flight. *No problem, I've done this before, like the day I flew in the lee of the Welsh mountains at 1200 feet coming back from Caernarfon and had turbulence all of my own making when four tornados flew underneath me at 250 feet...*

The ILS was already set up in the aircraft, I captured it by flying level and then descending on the glideslope, still in cloud, still in light turbulence.

I transmitted again,

"Aberdeen approach what speeds should I fly? Gold Charlie Delta."

They came back with,

"Golf Charlie Delta.... Standby," then after a minute or so, "Golf Charlie Delta, I have just been on the phone to a 737 pilot. He says fly 180 on approach, 130 at the threshold."

So that's what I did. When the speed came down below 200 I put a stage of flap out, gradually feeding out flap as the speed reduced.

At the point of putting the gear down we went below the glideslope... *Whoa!!! What's going on?*

As a light aircraft pilot if I'm doing a visual approach, I control speed using the stick and rate of descent using the power. If I am flying a known datum such as an ILS or level flight, I will control speed with power. I had reverted to a visual approach technique even though I was on the ILS, in cloud. *Come on Phil, get it together.* I moved the power levers forward what felt like a really long way (they are much longer than the bit you can see) and got back on the glideslope.

"Golf Charlie Delta set QNH xxxx."

"QNH xxxx set Golf Charlie Delta."

We came out of cloud at around 2500 feet.

"Golf Charlie Delta, report when you have the field in sight."

"Field in sight Golf Charlie Delta."

"Golf Charlie Delta cleared to land, surface wind calm."

"Cleared to land, Golf Charlie Delta."

I was flying fairly well apart from the little bit earlier below the glideslope. The aircraft was on the speed I wanted and height was within 10 or 20 feet of what I was aiming for, nothing different there to my usual flying.

As I reached about a 5 or 6 mile final, I flicked the transmit switch the other way and said, "Cabin Crew, this is the Captain, brace positions please."

Soon I was approaching the threshold, and like you do before a crash, your world goes into slow motion, each second taking about 20 seconds to elapse.

I thought. *This is a good stable approach. Can I think of any reason not to land? If we go around, I don't really know how to fly this thing, so there are risks there, but if I have to, I can climb straight ahead to 3000 feet or something and then get more radar vectors...*

Three greens, full flap... have I missed anything else important?

I couldn't think of any reason not to land.

We flew over the threshold, at a good height, on the centreline at 130, with each second taking 20 to elapse...

I looked outside, nothing inside the cockpit can help me now.

In my peripheral vision, the ground ever so slowly started to rise. With each rise, I added a little bit of back pressure and reduced a small amount of power.

Sometime later (like a couple of seconds), I was all the way back with the power. I don't know if the jet asked me to retard or not, I couldn't have heard it if it did...

The wheels touched with hardly a squeak, followed shortly by the nosewheel.

The nosewheel touching woke up my wife, who was curled up in a ball in the jump seat with her eyes shut waiting for the crash. She shouted,

"WE ARE DOWN?! WE ARE ACTUALLY DOWN?"

I now had a big problem, because I was in danger of throwing it all away.

My mission was to crash and possibly die within the airport boundary within reach of emergency services so that somebody might live. I didn't know if I could land it. I imagined I would drive the wheels into the runway not being familiar with the cockpit height above the runway. I hadn't thought about what I would do if I didn't crash!

Still in the 1 second takes 20 mode, I consider, *how I am going to stop this thing?*

I consider reverse thrust, but I can't risk it. If I don't select it correctly it's either a high-speed crash off the end of the runway or a high-speed crash off the side of the runway.

I decide to wait for 80 knots... It is never going to happen. I don't have time. At about 105 knots I start to lightly brake. At 80 I brake hard and pull back on the stick which lifts the nose a little off the runway.

We stop with about 100 metres to spare. The sim is paused, wife and the 737 captain pat me on the back, I have my head in my hands for about

30 seconds. After that he shows me some things, we do some circuits but I'm done, we don't crash, but I can't fly accurately to save my life.

I wrote more or less this story to Pilot magazine, I got a reply from the editor saying, "Bob is in Australia, I'll pass it on." I didn't receive a reply.

Analysis.

This was a challenge. How far could an average PPL get without any instruction or training?

If it was real there are a number of things I would have done differently:-

1. I would have left the autopilot on while getting a full brief from air traffic.
2. If Autoland was available, provided somebody could talk me through the settings, cross-check them, and how to know if it had failed and I needed to take control and go around, I could (should) have considered that if at all possible.
3. I didn't want to fly autopilot down to 200 feet and then take control of something totally new to land it. No way, that would have resulted in a crash for sure in my mind. I needed some time in manual flight to feel comfortable.
4. On the ground, I could have tried to get the engines into idle reverse, that would have helped with little or no risk, but without thinking about it beforehand there wasn't enough time.
5. I could have armed the lift dump and auto brakes.
6. From a trip on the jump seat (prior to 2001), I should have known to turn the engine ignitors on to prevent icing in cloud but I didn't. The engines kept running so this time I got away with it.
7. I should not accept anything outside my own personal limits, cloud base, visibility, crosswinds. If I had no choice, the risks would have increased, just like trying to use unfamiliar automatics would have been a risk increase too in my view.

Much of the above would have come out by correct briefing and planning before turning off the autopilot. Who knows, I might have even been able to stop on a high-speed taxiway instead of blocking the runway with burst tyres.

As it was, I'm sure I could have carried out the final actions:

1. APU – on – for cabin lighting.
2. "Aberdeen Tower, are we on fire? (overheated brakes). Do we need to evacuate?"
3. Parking brake – on.
4. Engines – off.
5. Go into the cabin and tell the cabin crew if we are evacuating – or waiting for stairs.

The overriding sensation from flying the 737 was two things:

Firstly you could feel the power of the engines making this 50 ton machine fly like it was weightless. It was an unstoppable force.

Secondly, she was an absolute lady in turbulence, especially compared to a light aircraft which would have been all over the place. What a magnificent machine!

In the 90's I looked at the Air 2000 jets (I even flew on one or two) and thought this was all beyond me, superhuman pilots flying super jets. I didn't have the confidence until I was in my 30's to even think about flying.

What do I think now? Age 60? Could I have done it? I think yes. Possibly I could. I could have flown a jet, but could I have operated it? Have I got the personality to find all the words needed to be a captain? I think probably not.

I am one of those people who knows they are in a simulator. I knew I wasn't going to die that day as a result of my actions in the Sim. Would I have been able to do it for real without panicking? I would like to think so, but I'll never know.

How would I behave in an emergency situation if it was real? Would I like to find out? Er no...

*

31 FLIGHT TO THE END OF THE WORLD

By Capt James McBride

Madrid Barajas International (LEMD) – Wednesday 3rd November – 1130am local time.

I sighed, *after all we've been through on this trip and now this?* We had just returned from southern Argentina, flying halfway round the globe and back, only to meet "Señor Jobsworth" on arrival in Madrid. The evidence was indisputable. The inspector was right and the proof was there on the documents in between us clear as day. My signature as pilot-in-command was on the bottom of the manual loadsheet for our Boeing 757. I was 'responsible' for any errors this legal document contained.

SAFA (Safety Assessment of Foreign Aircraft) Inspections can happen at any time on the ramp when you are flying commercially. Theoretically they are part of a program to ensure that operators maintain safety standards required by the EU. In practice, there appear to be more inspections of smaller operators than larger ones. Having been on the receiving end of several inspections, I was always ready to argue the case on behalf of the crew and our company.

"So, you see captain, it IS A FINDING!" The inspector's voice boomed out in the forward cabin of our VIP airliner. Looking around, I could see the crew were all busy working. There was much to do at the end of a VIP charter flight and none of them paid much attention to the inspector and his 'finding'. They were constantly moving through the cabin carrying equipment, gashbags, cleaning supplies and everything else needed to service our flying palace. I replied slowly,

"Hmm, well, the figures on the loadsheet may be in error, but not by much I think. Let's calculate it again shall we?"

I had drawn up the original loadsheet in the early hours of the morning in Cape Verde for our flight to Madrid having crossed the Atlantic. It was a relief to see the figures were as neat as they were.

"Plus, forty kilogrammes? Yes, okay I see it now, I was forty kilos over." I looked up and he was smiling.

"Yes, so you see captain I will record it as a safety assessment FINDING." I protested,

"But it was NOT unsafe Señor! Yes there was an error in the calculations, BUT it was on the SAFE side. I calculated the aircraft was forty kilos HEAVIER than the actual weight for takeoff. So, it was a GREATER margin of safety than we actually needed." He was writing on his clipboard. I knew I was not going to win this one. His face was set now as he said pedantically,

"If the loadsheet calculation is in error, then it is a FINDING." *Ugh, that word again...* I shrugged,

"You know Señor, in reality we cannot say exactly how much an airliner weighs to the nearest half-tonne! Anyway, you write up your report and I assure you I will personally take it to the CEO of our company for his actions."

Behind him, I saw one of our pilots shaking his head and smiling at this. I was the CEO.

*

It all started 5 days before. It was in Madrid that we began our flight to the end of the world. No joke. Our VIP charter with almost 50 executive passengers was headed towards southern Argentina. We planned to fly to Ushuaia, the southernmost city on the planet. The departure was scheduled for the evening and we would fly through the night. There had been a lot of last-minute arrangements which had to be made, but we did ferry the aircraft to Madrid at least 24 hours prior to the departure. You could say we were rested but stressed, if there is such a thing!

Although our private airliner crew were technically 'Off Duty' in the period leading up to the long night-flights, that does not tell the real story. As usual we were all heavily involved in provisioning the jet and preparing in

Madrid. That is the thing with a small company, everyone has at least two jobs. In my case, although I was 'CEO and Accountable Manager' (for the Civil Aviation Authority) I was also a training/check captain and commercial liaison with the broker. Oh, and of course I was a trolley pusher in the supermarket when the cabin crew loaded up with provisions.

Don't get me wrong, it was great fun all of it and I would not have missed it for the world. The best part was having this fabulous team around me who were so "GO-minded." Their positive mentality was infectious and together we felt like we could conquer the world or at least fly to the end of it!

Our client was Repsol, the Spanish oil giant which owned 98% of the Argentinian fuel entity called YPF. Repsol were sending a large group of their top executives to Argentina and back on a 5-day work mission. Given the time zone changes and the planned flight times, this was going to be an endurance test for us all. In addition to their work meetings they had planned, there were some sightseeing trips involved too. But the schedule was super-tight. Their only chance for proper rest was on the longer flight sectors. Like on the eight-and-a-half-hours from Madrid to Recife, Brazil across the Atlantic.

Fortunately, our VIP Boeing 757 cabin had been reconfigured with comfort in mind for long-haul flying. Normally a 757 would have 230 economy class seats, but ours was very different. In the forward cabin there were business class leather recliners set around hi-lo tables in 'club seating' configuration. The middle was fitted with forward facing recliners, while the aft cabin had 'club seating' round hi-lo tables again. A total of 80 passenger seats. In those days we did not have the bedroom onboard – that came later. It was a true all business-class airliner with fabulous cabin service to match.

We were operating 'heavy-crew', which meant we had two pairs of pilots, plus a travelling engineer (the flying spanner) and a few extra cabin crew. I always felt sorry for the cabin crew as they would do the lion's share of the work. Even though they could get some rest while we were flying, they would be on their feet for the best part of 19 hours on the outbound journey. Just the thought of walking across the Atlantic Ocean made me feel tired.

On the evening of our departure, I felt confident that we were ready in good time for the start of our epic mission. Fuelling had been completed, catering was being loaded, the crew were preparing the cabin while the on-duty pilots were in the cockpit getting ready for the first leg to Cape Verde. We had almost an hour to spare before the arrival of our VIP guests and we would be ready to receive them well before that. The ad hoc nature of VIP charter flights means that you have to give yourself as much time as possible to allow for any contingency. So let's wind the clock back and see where it all started.

*

Madrid Barajas International (LEMD) – Friday 29th October – 1900 local time.

Out on the ramp, I became aware of something not quite right. Standing at the top of the steps at the middle cabin door, I had a good view of the overall picture as the sun was going down. Our travelling engineer Sam had been doing his preflight checks under the left wing. I saw him walk purposefully towards Dominic the off-duty first officer. A few words then they both moved quickly towards the portside landing gear – *uh oh, I don't like the look of that.*

A few seconds later the pilot reappeared and looked up straight at me. By his expression I knew this was not good news.

"I think we have a puncture! The left outer mainwheel is flat." The engineer looked grim as he broke the news to me. He showed me the tyre pressure gauge – hardly any reading on it. Damn!

I looked at my watch. Only 45 minutes to pax boarding now.

"Can you get some air into it Sam? Talk to the Iberia airline engineers and get a pump." This was serious. A mainwheel change is not something to be undertaken lightly. Normally it is a job for the hangar, but here we were with only a short time before we were due to fly halfway round the planet, from a foreign airport and we had to act quickly. You see in VIP operations, delays due to technical problems are almost unacceptable to the clients. They assume that because they are 'flying private' that everything will be perfect. And oftentimes it is of course!

The pump was delivered and our flying spanner rigged it up. As fast as he was putting air in, it was coming out. Damn! Dominic and I were watching him with sinking hearts.

I am not an engineer, but I have seen enough aviation maintenance procedures down the years to know that changing a mainwheel on the ramp is not a five-minute job. Again, I looked at my watch, *ugh, this is going to be close.*

"We'll get the spare* out of the boot and I will talk to the handling agent to get Iberia to send a hydraulic jack, is there anything else you need?" Sam shook his head. Then to Dominic,

"Let's get moving matey, that mainwheel will take two of us to pull it out."

As we shinnied up the baggage belt into the rear hold of the 757, I smiled at the way we had all adapted to VIP airliner operations. You see in the normal airline world, this would be somebody else's job, but as mentioned before, we all had at least 3 of those. If we did not help the engineer, he would be unable to change the wheel, the charter would be late or cancelled and the company would suffer a huge loss. Such is the glamorous life of the Bizjet pilot.

We had just taken the straps off the mainwheel and were manoeuvring it out of the hold door onto the baggage conveyor when I became aware of somebody standing on the concrete down below.

"Captain! What is happening? Is the flight going to be delayed?" I saw a diminutive figure in high heels, a bright yellow mini-skirt and smart blazer with a cravat. She also had a loud voice for me to hear her so clearly above the cacophony of the ramp and the APU thundering away above our heads. I was about to meet Maria, the Spanish representative from the charter broker.

Her handshake was almost as strong as her voice as she introduced herself, giving me her business card.

"I am Maria. I would like to board the passengers in twenty minutes captain. Is that a problem for you?" Her tone was imperious and I thought then, *this is going to be a long five days*. I also thought about asking her why she was not wearing a high-visibility vest while out on the ramp, but then I realised her mini-skirt was visible enough. After all you could probably see it from space.

"No problem at all Maria," I smiled broadly, "we're just changing a wheel and then we'll be good to go." As if this was the most normal thing in the world. Another smile.

"The tyre is bad?"

"Just a little puncture, nothing too serious. We will change it for safety reasons." *Like we had an option? Hahaha.*

"Okay captain we start boarding in twenty minutes..." And with that she turned smartly on her stiletto and left. We rolled the mainwheel across the ramp under the fuselage to the left maingear to find the engineer cursing big time.

"They've only gone and brought the wrong fucking jack!" He indicated that it would not fit the crossbeam on the landing gear. He continued, "this is for a 737, jeez what's the matter with them?"

I had a quick conversation with the handling agent and was assured they would bring a Boeing 757 jack. But a few minutes later when the van arrived, it contained an Airbus hydraulic jack. I kid you not. Airbus A340 I think – anyway it would not fit. Our flying spanner looked very angry and I was worried about his blood pressure. I pleaded with the handling agent.

"You MUST make them understand, IT HAS TO BE A 757 JACK, for a Boeing 757. Iberia operate them for heaven's sake. In fact, they operate almost every aircraft type apart from Concorde!"

Soon after this while waiting for the next Iberia van to arrive, the bus arrived with the passengers and boarding started. Fortunately, our wonderful cabin crew had worked their magic and the interior looked splendid. Magazines, petit-fours, treats, departure drinks being served – I marvelled at the way they looked after our executive guests.

'Club four' seating in the forward cabin.

Meanwhile back at the comedy show. The van had returned with a very tired looking B757 jack and the engineer had installed it under the bogie. It was one of those with hydraulic pipes running from the pump in the back of the van. The pump was making a horrendous noise, but at least the aircraft was being lifted. It had started to rain. But we were making progress. I had reassured the pilots in the flightdeck that we were doing all we could to get the wheel change done and for them to delay the flightplan if required. The cabin crew were distracting the guests and all was good. We were going to do this.

It takes some time to get all those wheel nuts undone (there are 16 of them) and by now first officer Dominic had his sleeves rolled up and was helping with the spannering work. We had just gone past our scheduled departure time; the wheel was ready to be removed and the other one standing by. I saw a flash of bright yellow out of the corner of my eye. She was here again, with an umbrella.

"We are going to be LATE captain James!" Oh dear, I checked the time. She was right.

"Well as you can see Maria, we are close to fixing the problem and we will be on our way shortly I promise." Now the wheel was off the axle, but to my horror I saw the axle was getting lower. A small rivulet of shiny hydraulic fluid was leaking across the concrete. The whole bogie was sinking – aargh!

"QUICK! Get the pump running again!" Shouted the FO as Sam ran to the back of the van. Again, the awful mechanical engine grinding sound came as they manoeuvered the new wheel into position. I literally held my breath as I witnessed what could be our worst possible outcome.

And yet... Yes! It had stopped descending. Now it was slowly coming back up. But the rate of hydraulic leakage was worse, *let's hope there's enough fluid in the pump reservoir to get this done...*

There was plenty of grunting and cursing as they got the wheel back on the axle and started torquing up the nuts again. I thought Sam was going to have a heart attack. It was just in time. I turned to Maria as calmly as

I could.

"We will be ready in the next ten minutes or so and we can make up the time on the Atlantic. Don't worry Maria, we won't be late on arrival in Bariloche."

With disaster narrowly avoided, we loaded up the old wheel in the hold and packed the tools away as quickly as possible. The Iberia engineer just looked blank as we gave him our candid opinion of their Boeing 757 hydraulic jack. I made a mental note to follow up later.

*

Soon we were climbing out of Madrid and settling into our leather business class recliners. It had been a close shave, but we had made it. I mentally composed my email to Iberia engineering head of line operations. Our crew rest area for these flights was in the rear cabin where we had screened off one of the club four sections with curtains. Inside it was dark and quiet, although we did not have bunks, it was conducive to sleeping anyway. As the first flight was five hours, there was plenty of time to relax.

After he had cleaned himself up, Dominic came to join me. I congratulated him on his fabulous effort assisting our engineer. I had no doubt we would have still been there on the ground without him.

The landing in Cape Verde was uneventful. We refuelled and the same pilots continued to fly the next six-hour leg to Recife in Brazil. Enroute I knew we would be crossing the famous ITCZ (Iner Tropical Convergence Zone) so we expected some bad weather on this flight. Sure enough, a few hours after getting airborne we could feel plenty of turbulence in the darkened cabin. When I looked out of the window from our rest area, I could see flashes of lightning from the thunderstorms way down below.

I was interested to know what was happening up at the sharp end. Being careful not to disturb the operating crew too much, I entered and took a seat quietly on the jumpseat behind them. Both pilots had their heads forward, peering intently out of the front windshield and the cockpit lighting had been turned up. They were actively avoiding the worst of the build-ups which they could see on the weather radar.

Avoiding a large active thunderstorm on the ITCZ.

Suddenly there was a flash of St Elmo's fire across the captain's window which arced across the glass. Although not in itself dangerous, it is indicative of close proximity to thunderstorms and gets your full attention. Legend has it that if you extend your index finger close to the glass, there will be a spark across to your hand. I have never been brave enough to try it. Nor has anyone else I know. When you are flying close to 40,000 feet at 500mph in the dark over the ocean, I feel it is unwise to experiment with nature.

St. Elmo's fire - dramatic at night.

*

Recife/Guararapes International (SBRF) – Saturday 30th October – 0300 local time.

It had been a relief to leave the bad weather behind us. The descent and landing in Recife had been smoothly executed. It was now our turn to fly.

Dominic and I would be at the controls for the third and final sector to Bariloche, Argentina. As they handed over the jet to us, I could see our two colleagues were exhausted. They had been on duty for a very long time, now they could rest during the next seven-hour flight. If you look at the Google Earth Image, it shows our progress across the surface of the planet.

It felt good as we raced down the 3300-metre main runway in Recife at three o'clock in the morning. The Boeing's controls came to life and once again I marvelled at the beauty of this machine. The 757 has a surplus of power and with it, a versatility which is uncommon in airliner design. A replacement for the 727, I like to think it was designed by engineers who still drove V8 engined automobiles to work in Seattle. It was the last airliner to be drawn up without reliance on computer-aided design. The airliner used only 30% CAD drawings during the process. The later B777 and 787 were produced almost entirely by computers.

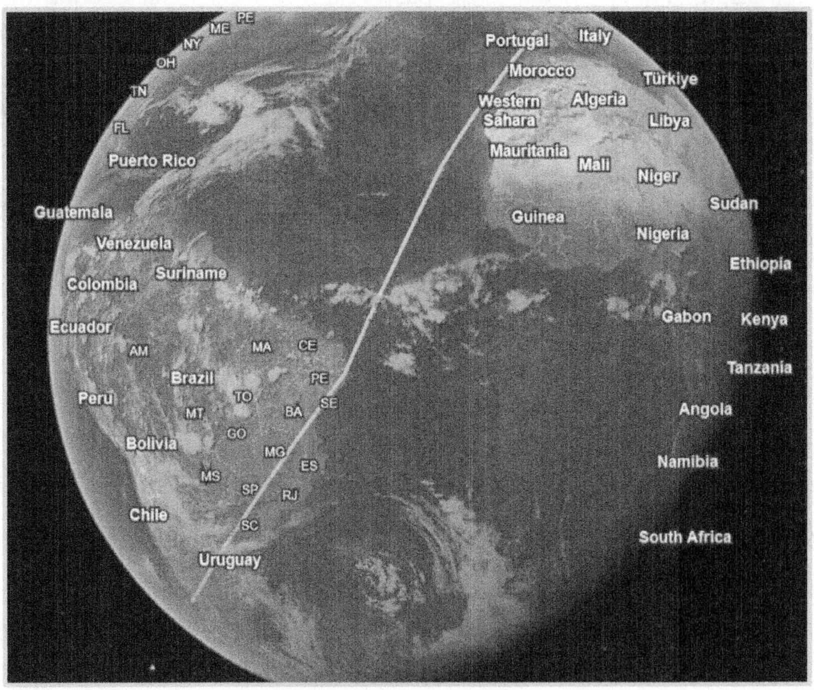

Our "Flight to the End of the World" - 11,400kms in a night and a day. Image from Google Earth.

As dawn came up we had magnificent views of the eastern seaboard of South America. On the righthand side, Dominic had a grandstand seat for the vast Brazilian rainforest, while I could see the giant Atlantic Ocean on the left. Breakfast came and went, the air was smooth and we were on schedule cruising sedately at close to Mach .80, we were very happy.

Lunch was served about an hour before top of descent and we had great views of Montevideo and Buenos Aires with little cloud cover to obscure the River Plate estuary. Our total flight time was seven hours and twenty minutes and we landed in Bariloche, the "centre of adventure tourism and chocolate capital of Argentina" – almost 20 hours after leaving Madrid. When we took stock later and looked at the map of the world, we were in the bottom righthand corner of the globe. The mission was at the limit of our range, but with two complete sets of pilots it had been possible and safe.

After disembarking our executive passengers and cleaning/securing the aircraft, our stay at the five-star El Casco Art Hotel made the journey worthwhile. The accommodation and catering were very fine indeed. Being surrounded by paintings and sculptures was a nice change from our usual airport hotels. Dominic and I were both enamoured with two large heavy bronzes standing by the fireplace in the lobby. They were full-sized hunting dogs and beautiful. We thought about it and realised it would be easy enough to bring them back with us in the hold of the 757, one each. Aware that all the art was available for purchase, we asked the price at the desk. $7000... each! That would have been an expensive souvenir, we declined.

*

San Carlos de Bariloche International (SAZS) – Sunday 31st October – 0800 local time.

Our trip today was an out and back mission from Bariloche to Puerto Madryn on the Atlantic coast with our large party of execs. They were booked to go Whale Watching. We had a couple of issues still to be solved. One of these was that the ATC at Puerto Madryn were Spanish speaking only – it was a domestic airport with a tiny ramp and even smaller terminal building. They had very recently resurfaced their runway so we ensured it would take the weight of our 757.

Finally, their Sunday opening hours were not confirmed. It seemed that even though our client wanted to depart from Bariloche at 9am, the air traffic services might not be available until 10am. Oh, and there was no radar guidance. Just a visual control room (VCR) with a visual approach to land – thank heavens the weather forecast was perfect.

The first problem had an easy solution. We would carry the handling agent on the jumpseat in the flightdeck wearing a headset. He would translate ATC instructions for us. The ground manoeuvring would be fine so long as we were cautious, especially with the thrust from our RB211s close to the single storey terminal. Blowing their roof off would probably be considered rude.

Our real 'stopper' was the opening hours for ATC. Unlike landing at a farm strip with a small aircraft, bringing the 757 back to earth requires some ground control personnel and a licenced airfield. On these flights, although flying as captain, I would be sitting in the right-hand seat. The pilot in the lefthand seat was Jerry, a captain under training. He was sitting there patiently waiting for our departure clearance. We still had half an hour to scheduled time of departure but... When I called the tower controller, he replied there was no flightplan in the system for us to fly to Puerto Madryn. Ugh!

I called our operations centre back in Athens; the phone conversation went something like this.

"Hi there, it's James in Bariloche, Argentina. We need the flightplan filing for our trip to Puerto Madryn please."

"Ahh! We are glad you called captain. The Puerto Madryn airport is not open, the airport has not replied to our messages, so you cannot depart." At this precise moment, Maria the broker rep appeared in the open flightdeck doorway, *yikes that woman has a nose for trouble!*

I held up my hand as she started to speak, indicating I was on the phone. No point, she continued anyway,

"Captain James, the passengers are wanting to know why we are still here. What is the problem?"

"Maria, I will be with you very shortly. I am on the phone now trying to resolve the issue, please assure the guests we will be going soon." She returned to the cabin and I spoke with our man in Athens,

"Look my friend, we have done our due diligence and I am sure the airport will be open for business as we arrive overhead at 10 o'clock local time." He was still not happy. He was the flight despatcher and he needed confirmation in writing from somebody. I tried again,

"But it is only a one-hour flight from here, we are carrying round trip fuel, we could easily come back if it is closed." Still no movement on his side, He was adamant and in fairness he was just doing his job. I asked him to email them again and said I would call him back.

My friend Jerry in the other seat shrugged his shoulders. I got it. He was an airline pilot through and through. In the world he operated, there was always somebody else to make things happen, while in the private jet arena, frequently the crew is left unsupported. It was time to do something. The aircraft was ready to go, all passengers were onboard and our lovely broker rep was chomping at the bit. Our despatcher needed something in writing.

Our Bariloche handling agent was already on the jumpseat in the flightdeck with us.

"Juan. Can you be certain the airport will open at 10 o'clock today exactly?"

"Si, Si, mi Capitán. It WILL open; I spoke to them yesterday. I am 100 per cent sure."

"Okay, let's go to your office, we need to do it now!" I smiled at Jerry,

"Hold the fort mate, I'll be back soon."

And with that we left. Fortunately, the air bridge was still attached to the forward door. Once out of sight of the passengers we started running.

"RAPIDO Juan! Let's GO!" I shouted and he moved himself faster than I thought possible for a man of his years. As soon as we got to his office, I got on his desktop computer and started typing...

We were back on the jet less than 15 minutes later and as we arrived breathlessly back in the cockpit, the mobile phone was ringing. It was Athens. He sounded excited.

"CAPTAIN! We have had an EMAIL from Juan _______ the handling agent in Bariloche! He has spoken to Puerto Madryn and HE GUARANTEES the airport will be open for your arrival. Your flightplan is in the system now."

"Oh! ... thank you my friend... that is very good news. We will call for start-up with the tower now. Talk later, thanks again buddy." Juan was getting his breath back on the jumpseat, Jerry was shaking his head grinning and I called the tower to get our departure clearance. *Just breathe...*

*

Puerto Madryn (SAVY) - 5000 ft, 210kts, overhead. Sunday 31st October – 1000 local time.

I was operating the radio and Jerry was flying the jet. He was an experienced 757 aviator with plenty of time on type, so it was a joy to fly with him. In addition, he had a great sense of humour and we got on very well together – always a bonus. I started calling Puerto Madryn tower on the radio – no answer. By arrangement with the radar controller, we had begun a holding pattern in clear sky at 5000 feet above the airport. It was a beautiful sight, with the town just on the coast and the new runway standing out well just a little inland.

"Juan, you need to call them in Spanish. Here's the transmit button." I pointed at the radio panel next to him.

To my delight they answered immediately and Juan translated for us,

"You are cleared to land capitán."

"Thanks Juan, tell them we will make our approach now." Looking across to my left, Jerry looked puzzled. He had heard the radio conversation and my interaction with Juan, but he was troubled.

"Whassup?!" I asked him smiling. "We're cleared to land mate."

"Well... what should we do now? I mean, how do we get down from here?" I looked across at him. *Maybe another joke? No, he's not joking it's a serious question.*

Of course! In the airline world, airliners run on rails in the sky. Published arrival and departure procedures, everything preordained, planned and executed. Here we were in the boondocks. There were no procedures, we would have to wing it.

"Just head out to sea, descending, then bring it back in on a five-mile final at 1500 feet or so. That will work out well for you." Which is exactly what we did. I put the extended centreline of the runway in the computer for him so he had something on the moving map display to aim at, but otherwise it was a visual approach just like in a Cessna. He kept the autopilot in all the way round, which I thought was a little overcautious, but I was okay with that.

We did make a radio call again on short final. Juan made the transmission, but there was no response. We had been cleared, so we landed. After landing Juan asked the tower for instructions. He told us,

"Backtrack the runway to exit left onto the apron." There was little else. Because we were so lightweight we had finished our landing halfway down the runway. The expeditious way was a 180 degree turn in our present position then and backtrack. Jerry slowed the aircraft down to a walking pace and brought it close to the righthand side of the concrete. Looking over his shoulder, he wound on full left steering angle with the tiller on his side and increased power on the righthand engine to assist.

Once back on the centreline we were taxying slowly towards the apron exit. We heard some Spanish on the radio with our callsign and Juan replied. I looked back at him with my hands open, *what now?*

"Capitán, he says you should have gone to the end of the runway to turn."

"Well it's a bit late now Juan, we've done it. We will talk to them later."

Jerry taxied very carefully onto the concrete apron leaving as much space as possible on the other side, but there was a lot of 757 everywhere. From our flightdeck we looked down on the terminal building.

Maria seemed pleased, although the cabin crew were all complaining to me about her. According to them, she had been interfering and criticising the cabin service. I did my best to pour oil on trouble waters as we disembarked the passengers. We would be waiting here for five hours while they enjoyed their whale watching tour. Then we would fly them back to Bariloche.

Once the cabin was empty, we could relax. There was no handling agent, so we asked Juan to liaise with the airport for facilities. The crew needed to wash and dry all the serving dishes, crockery and cutlery. No small task after 50 guests having been served a full breakfast enroute. The airport terminal café had a kitchen with sinks, so the cabin crew set about offloading bar trolleys full of dirty dishware.

A representative from the airport then appeared.

"Capitán, you have damaged our new runway surface because you turned 180 degrees in the middle of the runway, not at the end." I had checked that the PCN (Pavement Classification Number) was high enough for the B757 to manoeuvre, so I did not believe we could have damaged it, but...

"I want to see the damage my friend; we will go now." And he led the way down the steps onto the bright concrete of the parking apron. We jumped into the operations pickup truck, along with a couple of other airport workers and drove out to the runway. He received the Spanish radio clearance to carry out a runway inspection and we accelerated along the centreline. About 100 metres away, I saw a black lump on the tarmac which matched the one in my throat. Surely not? That was not there when we landed...

As we got closer, the lump became clear. It was a spider, as big I have ever seen. He stopped the truck and we got out.

"Wow! Look at the size of this thing!" I took photos of it, but did not get too close. The ramp airport ops supervisor was amused at my reaction.

"You can take it home with you if you like." He offered.

"No chance of that my friend, I'll leave it here. Now where's this damage to the runway surface?" Another 200 metres further on, we got out again. Here were our tyre tracks as we completed a minimum radius turn, black rubber was evident on the tarmac surface. I got down close and checked, urging him to do the same.

"Feel with your hand," I motioned, "it is smooth. No damage to the surface, just some rubber from our tyres and I won't charge the airport for that!" He shrugged, but he had to agree with me.

After more photos (to be used as evidence in the future should I need it), we returned to the terminal. I was amused to see that our Boeing 757 dwarfed the buildings there.

Now the airport supervisor came to see us, he was accompanied by Juan to translate.

"Another problem capitán. They want you to move the aircraft to give more space on the apron. A scheduled flight is coming in one hour's time and there is not enough room for that airplane to park."

"Please ask the supervisor how he expects us to move our aircraft, does he have a tractor with a towbar which would fit?" A pause while this was translated into Spanish. It was clear the answer was no, which is what I expected. I grimaced and then motioned for them to follow me. We walked across the ramp and they told me how much space the Regional Jet needed. We paced it out; *hmm this might just work.*

"If we could move the 757 back to HERE," I drew an imaginary line on the ground with my foot, "they would have enough space right?" Both nodded their heads in agreement. It would involve repositioning our flying palace back about 40 feet from its present location. I looked at the right-angled taxiway out to the runway, not enough room to manoeuvre there for sure. There was always the option of taxying out after the regional flight had landed, holding on the runway, then coming back in after they had left.

I was not keen on that idea either. There were all sorts of risks involved, not least in the communication department as we had already witnessed firsthand. The most logical answer was a powerback. When I mentioned this to Jerry, he shook his head in wonderment, but I assured him it would be a relatively easy solution. I had done a powerback once before with a 757 and that had been with a fully laden aircraft in Agra, India. This time with a lightweight machine, a clean ramp and even a marshaller I could trust, it would be a cinch.

We briefed the ramp team and Sam would be our marshaller as we reversed the 757 towards the western edge of the ramp. We also informed ATC of our intentions – they were pleased that we would be resolving the ramp congestion issue before it occurred. I warned everyone about the noise. The Agra powerback had been thunderous, they must have heard it at the Taj Mahal.

After all the holds and doors were closed, Jerry and I sat in the flightdeck with Juan. We kept the comms simple.

"Tell them we are requesting engine start up to powerback please Juan." Not sure how that translated into Spanish, but the ATC tower was expecting it anyway.

We started the engines then gave the thumbs up to our engineer standing well off to the side of the ramp. He began to wave us back, so I selected reverse idle and released the parking brake. The 757 began to move almost immediately, rolling slowly backwards over the smooth concrete. I had both feet flat on the cockpit floor and had already briefed Jerry to stop me if I tried to go for the toebrakes. Sitting the airliner on its tail would create a lot more problems.

As Sam crossed his arms above his head, I deselected reverse and there was a slight pause as the reverser cowlings closed, we kept rolling backwards! Then I added a little forward thrust lever movement and we stopped. I applied the parking brake and we shut down the engines. *Hmmm that might have been closer to the edge of the ramp than we wanted.*

As we came down the steps Sam was smiling at us,

"I thought you weren't going to stop then Boss!"

"Hahaha! Neither did we."

I was relieved when I looked back behind the engines. Our positioning was perfect. The tail was hanging out over the scrubland off the concrete and now there was plenty of room for the RJ.

*

I noticed for the first time a YPF fuel truck not far away and wondered if it was coincidental.

Soon the cabin crew reappeared with all the clean crockery and serving utensils. We were ready to go again. Shortly after that, the RJ made its landing and taxied slowly onto the ramp. The pilots were surprised to see us. After disembarking their passengers, they came to visit the 757 and had a look in the flightdeck. They were impressed, not only with the beauty of the interior, but the roominess of the flightdeck is something to behold. It is actually wider than a Boeing 747 cockpit, with fabulous big picture windows. Several handshakes and "Muchas Graciases" later they departed and I suspected there would be two more applicants for Aerolineas Argentinas within the next week.

In the mid-afternoon our customers arrived back from their whale watching trip and apparently it was a great success. Nobody noticed we had moved position on the ramp, not even Maria. After gracious thanks to the ramp and ATC people of the airport we made our departure. The return flight was an anticlimax thankfully. The only 'fly in the ointment' was Maria, who once again upset the cabin crew with her constant interference while they tried to deliver their five-star service to the passengers..

Sitting the airliner on its tail would have provided more problems.

The following day the other flightdeck crew operated to Ushuaia and back for our guests to see the Beagle Channel and penguins. The passengers seemed to get even less rest than we did. During the flights they would all be awake, reading, writing, on their laptops and Blackberrys. Or they would be having active discussions with their co-workers.

We understood before and after the flights in Argentina, they were also busy meeting with the senior people from YPF. It was an impressive performance by a highly motivated team of executive managers.

*

San Carlos de Bariloche International (SAZS) – Tuesday 2nd November – 0800 local time.

Checking out of the hotel early, we were ready for passengers well in advance. Again, I had to hear the complaints from the cabin crew about "interfering Maria." If their reports were to be believed, the charter Rep had exceeded her authority by a significant margin during the out and back flights to Ushuaia on the previous day. Although I had been loathe to upset the status quo on previous days, we would now be making our way home from this point. I knew that we could manage without Maria's help from here on if we needed to. It was time for somebody to act.

It was normal practice for us to help with baggage handling while operating the VIP 757. One of the pilots and usually the travelling engineer would be helping load and unload the cargo holds down below. I quietly asked Jerry to identify Maria's checked bag and put it to one side. He placed it on the ground outside, visible from the flightdeck. When Maria arrived onboard, she immediately went to the galley and started to boss the cabin crew about, I interrupted her.

"Maria, could we have a word in the flightdeck please? Thank you." All eyes turned towards me,

"And yes, carry on ladies please, carry on. Passengers will be boarding shortly. You know your duties."

In the quiet of the flightdeck, there was just Dominic who was busy completing the paperwork. He pretended to be even busier as we entered and I asked her to sit down on the spare jumpseat behind him. I closed the cockpit door and took my seat, turning to look at her. She opened her mouth to speak.

"No Maria. Please listen to me now, we do not have much time. Forgive me for being so direct, but you have a decision to make. If you wish to travel with us today and tomorrow back to Madrid, then you will do as I say." She looked irritated, then annoyed, then confused and finally when I had finished speaking she looked resigned. My final words were along the lines of,

"That," I pointed outside the window, "is your suitcase, am I right?" She nodded, her eyes burning brightly.

"I tell you now, as commander of this aircraft, if you interfere with any member of my crew from this moment, I will leave you and your bag here in Bariloche. Do I make myself clear?" She nodded her head again,

"Yes but captain..." I held up my hand,

"No. No more words Maria. You have said enough. Your interference with my crew is a flight safety hazard we can do without. If you need to speak to anyone, you come to me, that's it!"

As she went out closing the door behind her, Dominic turned round with a huge grin on his face. Then the door opened again and it was the senior cabin crew, our number one. She looked delighted, it was written all over her face.

"AND YOU!" I said with mock severity, "should not listen at the flightdeck door!" She burst out laughing.

"Ooh yes my captain, whatever you say my captain." I couldn't stop myself smiling.

"Coffee please and board the passengers when you're ready."

"Ooh yes my captain, whatever you say my captain..."

The rest of the day went well. We flew up to Iguazu in Brazil and landed there. Our passengers went sightseeing one last time to see the famous Iguazu Falls. On the way to the airport prior to landing we negotiated an aerial flypast of the waterfalls at quite a low altitude with ATC. They were very helpful and we got the impression it was not an unusual request.

Maria did not say another word to anyone, apart from a very stilted farewell to me in Madrid. She ignored our wonderful cabin crew for the flights home and yes we did remember to put her bag in the hold. It went on last of course and came off first at destination. The passengers were very complimentary about the service and the flights.

It was with a great sense of achievement that we taxied onto stand at Madrid, only five days after we had left. We were glowing at the fact we had flown to the end of the world and back. Even Señor Jobsworth the SAFA Inspector could not ruin our good mood – although he did try.

The best part was having this fabulous team around me. VIP Boeing 757 crew.

*We had learned the hard way that a "Flyaway Kit" was essential equipment when using airliners like VIP jets. Our company had suffered a punctured nosewheel on a flight to Africa, which resulted in a spare wheel being flown down from Europe in a Learjet! So now we always carried one of each, a nose and a mainwheel strapped down in the read hold. After the Madrid debacle, we also carried the correct jack.

**A powerback in a Boeing 757 is a very rare manoeuvre. I recall it was in the Flight Crew Training Manual back in 1989 when I first completed my conversion course on the type with Air 2000. Eastern Airlines in Miami had plenty of 757s and I am sure they will have done a few powerbacks. They certainly used to powerback the MD80 series of airliners, but they had engines high up at the tail. I also remember my senior colleagues (all captains) discussing it and saying how they never ever planned to do one. Nobody wants to be the one to sit an airliner on its tail. Which of course is one of the major hazards

They call Ushuaia in Argentina "The end of the World" – it is the southernmost city on the planet.

*

Designed by guys who drove V8 engined cars to work. B757 an aviation Icon.

Definitely not a Ryanair ship. The VIP B757.

Glossary of Terms and Abbreviations

AAIB – Air Accident Investigation Branch (of Dept for Transport UK)

ABP – Able Bodied Passenger

ACMI – Aircraft, Crew, Maintenance and Insurance – also known as "Wet Lease" (i.e. WITH crew)

ADD – Acceptable Deferred Defect (also called "Hold Item")

ADF – Auto Direction Finder unit – often associated with NDB see below

AEW – Airborne Early Warning

AFDS – Automatic Flight Director System

AGL – Above Ground Level

APU – Auxiliary Power Unit

AOC – Air Operators Certificate

ASR – Air Safety Report

ASI – Air Speed Indicator

ATA – Actual Time of Arrival

ATC – Air Traffic Control
ATC Slot – See Slot

ATD – Actual Time of Departure

ATIS – Automated Terminal Information System gives aerodrome current weather

ATL – Aircraft Technical Log or "the Techlog" (RAF Form 700)

ATO – Approved Training Organisation

Autobrake – as it says, an automatic braking system for landing

A/C – shorthand for Aircraft, can also be seen as AC or a/c.

BAe – British Aerospace

Baro – Barometric Altimeter

Blocks – as in "Off Blocks" – literally means the chocks holding the wheels – same as "Off Chocks"

BER – Beyond Economic Repair

BSI – Borescope Inspection of engine

CAA – Civil Aviation Authority

CAPT – Captain or Aircraft Commander

CAS – Calibrated Air Speed = IAS corrected for instrument and position error

CAVOK – Cloud And Visibility OK – very nice weather

CBT – Computer Based Training

CDL – Configuration Deviation List

CDU – Computer Display Unit

CFI – Chief Flight Instructor

CFP – Computer Flight Plan

CFIT – Controlled Flight Into Terrain

CGI – Chief Ground Instructor

Chocks – large blocks which are stuck next to the wheels on the ramp – stops the aircraft moving if the parking brake fails

CIP – Commercially Important Passenger – same as VIP

CLB – Climb abbreviation for FMC

Coffin Corner – as the name implies, a rather undesirable place – the small corner of the flight envelope between slow speed stalling speed and high- speed buffet speed

CONFIG – Configuration of the aircraft, flaps, gear, speedbrakes etc.

CONFIG Warning – warning that configuration is incorrect for manoeuvre

CPT - Captain

CRM – Crew Resource Management

CRMI – Crew Resource Management Instructor (a facilitator really)

CRS – Certificate of Release to Service (maintenance term)

CRZ – Cruise abbreviation for FMC

CRZ ALT – Cruise Altitude abbreviation

for FMC

CSR – Cabin Safety Report

CTBL – Contactable – referring to a crew member being available to call by crewing

CTC – Chief Training Captain (usually Postholder for Training on behalf of CAA)

CTOT – Calculated TakeOff Time – used on Flightplans

CVR – Cockpit Voice Recorder

Cyan – Colour of blue specific to Boeing EFIS displays

Damp Lease – an ACMI agreement with cockpit crew only

DDG – Dispatch Deviation Guide

DEC – Direct Entry Captain

DFDR – Digital Flight Data Recorder

DFO – Director Flight Operations

DOB – Death On Board

DODAR – Diagnose, Options, Decision, Action, Review

DV Window – Direct Vision window in flightdeck which slides open on ground

EAT – Estimated Approach Time (normally for holding aircraft in the stack)

EADI – Electronic Attitude Director Unit (part of EFIS instrumentation in Boeings)

EET – Estimated Elapsed Time (enroute on a Flight Plan)

EFIS – Electronic Flight Instrumentation System

EFATO – Engine Failure After Takeoff

EFOTO – Engine Failure On Takeoff

EHSI – Electronic Horizontal Situation Indicator

EMA – East Midlands Airport (IATA Code)

EOBT – Estimated Off Blocks Time

EPR – Engine Pressure Ratio – an expression of how much thrust is being produced

ETA - Estimated Time of Arrival

ETOPS – Extended Range Twin Engined Operations – often over water (e.g. Atlantic)

FCOM – Flight Crew Operations Manual

FCTM – Flight Crew Training Manual

FCU - Fuel Control Unit

FDM – Flight Data Monitoring

FDR – Flight Data Recorder

FFS – Full Flight Simulator

Final Line Check – After Line Training is complete to release crew member to Line

Fire Handle – shuts off fuel, hydraulics and when rotated activates Fire Extinguisher

Flightdeck – the little room at the front where the pilots sit

Flight Level – or FL – an altimeter indication in 000's of feet referenced to 1013mb

FMC – Flight Management
Computer

FO – First Officer also known
as Co-Pilot

FOD – Foreign Object Debris

FOM – Flight Operations Manager – usually Postholder reporting to the CAA

FOM – Flight Operations Manual

FPL – Flightplan

FSTD – Flight Simulator Training Device

FSO – Flight Safety Officer

FTL – Flight Time Limitations

G/A – Go-Around, aborted landing. (Used to be known as Overshoot)

Glass Cockpit – refers to the introduction of computer screens in flightdecks EFIS

GPU – Ground Power Unit – provides electrics to run the aircraft without APU

GPWS – Ground Proximity Warning System (does what it says on the tin)

GRADE – Gather (info), Review, Analyse, Decide (& Do), Evaluate

Ground Lock Pins – heavy duty bolts with flags to lock landing gear down

HF – High Frequency – Military Radio Network for Military A/C

HSI – Horizontal Situation Indicator (Orville and Wilbur called it a compass) Hold Item – similar to ADD. Holding over until repair is possible

HOTAC – Airline speak for "Hotel Accommodation"

HOTAS – Hands On Throttle And Stick – buttons for vital functions placed on these controls

IAS – Indicated Air Speed

IFALPA = International Federation of Air Line Pilots' Associations.

IFR – Instrument Flight Rules

ILS – Instrument Landing System – for precision approaches to runway in IMC

IMC – Instrument Meteorological Conditions – cloudy outside the cockpit windows

IR – Instrument Rating – a licence to fly aircraft on instruments alone

IRS – Inertial Reference System

IRU – Inertial Reference Unit

ITCZ – Inter Tropical Convergence Zone

LCA – Low-Cost Airline

LCC – Low-Cost Company

LCZR – Localizer (element of the Instrument Landing System)

Line Check – annual event for all crew

Line Training – after basic/initial training in Simulator, training on the line

LLZ – Localizer (azimuth) element of the ILS

LNAV – Lateral Navigation system – part of the AFDS

LOFT – Line Oriented Flying Training

LPC – Licence Proficiency Check (for pilots)

LST – Licence Skills Test (for the issue of a Type Rating)

LTN – Luton Airport (3 letter IATA code)

LVPs – Low Visibility Procedures

LVOs – Low Visibility Operations

Mach No – associated with cruise speed of airliner as percentage of Mach 1 – the local speed of sound. For example 'Mach 0.8' = 80% of Mach 1.

Magenta – purple type colour specific to Boeing EFIS system MAN – Manchester Airport (3 letter IATA code)

MEL – Minimum Equipment List

METAR – Meteorological Actual Report – "the latest weather"

MPA – Maximum Power Assurance engine runs – Engineering Technical MPL – Multi Pilot Licence

MOR – Mandatory Occurrence Report – to CAA (or regulating authority) MSA – Minimum Safe Altitude

MSD – Minimum Separation Distance (used by the Military to describe closest distance to ground or other surface obstacles)

N1 – engine gauge showing speed of front fan of large bypass engines N2 – second stage fan

N3 – third stage fan

NDB – Non Directional Beacon – used with aircraft auto-direction finder

NPA – Non-Precision Approach on instruments to an

airfield/aerodrome

OAT – Outside Air Temperature – also known as Ambient OPC – Operator Proficiency Check for crew

P1 – Senior Pilot onboard – another name for Captain – refers to logging the flight-time

P1/S – First Pilot under Supervision/Training P2 – Co-pilot

PA – Public Address system – aircraft loudspeaker system to cabin

PF – Pilot Flying (also known as the Handling Pilot – with hands on controls)

PIC – Pilot In Command

PAP – Passenger x 1 PAX – Passengers Plural

PIREP – Pilot Report – usually of meteorological phenomena

PLI – Pitch Limit Indicators

PLOC – Prolonged Loss Of Communication (on the radio) – due to losing contact with ATC

Pusser – Generic Term for Royal Naval Supply Branch

QFI – Qualified Flying Instructor

QRH – Quick Reference Handbook – with emergency & non-normal checklists Quarter-Mil – a topographical Map with scale 1:250,000

QWI – Qualified Weapons Instructor

RA – Radio Altimeter

RADALT – Radio Altimeter

RFDS – Royal Flying Doctor Service

Rotate – Call by Monitoring Pilot to indicate to PF that it is speed to get airborne

RPM – Revolutions Per Minute – with reference to engine speed usually

R/T – Radio Telephony – often abbreviated – VHF in Civil World

RTO – Rejected Takeoff

SAS – Special Air Services

SBY – Standby – referring to crew usually associated with rostering/crewing dept.

SCCM – Senior Cabin Crew Member

SID – Standard Instrument Departure – defined for all flights heading that direction

Slot – An ATC timed departure restriction – usually with an allowance of -5 and +10 minutes

SMS – Safety Management System

SOP – Standard Operating Procedure

"Soup Dragon" – (slang) Cabin Crew in forward galley serving flightdeck

SP – Safety Pilot – qualified observer on jumpseat in flightdeck for Line Training

Speedbird – British Airways' radio callsign

SSA – Sector Safe Altitude – usually within 25 nm

STA – Scheduled Time of Arrival

STAR – Standard Arrival Routing – defined for all flights from that direction STD – Scheduled Time of Departure

TAS – True Air Speed = speed of a/c relative to the airmass

TCAS – Traffic Collision Avoidance System

Techlog – aircraft Technical Log; a legal document – see also ATL

Thrust Reverser – on each engine deflects exhaust forward to reduce speed on landing

Topple-Free – Gyro Attitude Indicator immune to toppling

TRE – Type Rating Examiner, check airman, usually a Training Captain

TRI – Type Rating Instructor, usually a Training Captain

TRTO – Type Rating Training Organisation

TWI – Tactical Weapons Instructor

TWU – Tactical Weapons Unit (training pilots to fly with weapons)

UNMIN – Unaccompanied Minor – an escorted child passenger with a large tag on a lanyard round his/her neck

UTC – Universal Time Coordinated (used to be known correctly as GMT)

V1 – Decision Speed on takeoff roll beyond which we must fly

VFR – Visual Flight Rules

VHF – Very High Frequency radio – civil airliners use it

VIP – Very Important Person – many Bizjet pax are VIP (see CIP also)

VMC – Visual Meteorological Conditions

VNAV – Vertical Navigation System of FMC for automatic flight

VNE – Velocity Never Exceed

VOLMET – VHF Met reports automatically broadcast – Originally from the French VOL = Flight & MET = Meteo

VOR – VHF Omnidirectional Radio Beacon

VSI – Vertical Speed Indicator – in feet per minute for climb or descent

VVIP – Usually applies to Head of State, President, Royalty or Head of Government (Prime Minister for example)

Wet Lease – an aircraft leased to another operator
WITH crew

WTL – Worn To Limits – engineering technical

*

THE FLIGHTDECK SURVIVAL MANUAL

How to survive a career flying aeroplanes for a living.

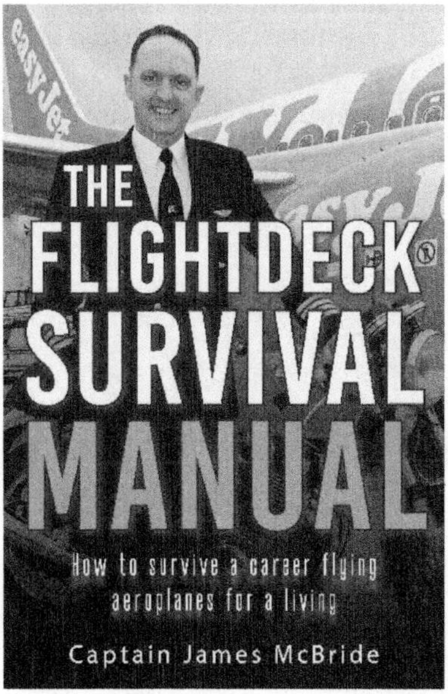

Dive into the captivating world of aviation through the eyes of an experienced pilot who expertly narrates the trials and triumphs of modern airline crews.

With knowledge amassed over a 40-year career, these stories illuminate the technical, environmental, and human factors that shape today's airline industry.

Each narrative invites readers to step with crew into the cockpit or cabin to navigate unforeseen challenges that can alter the course of a flight in an instant.

McBride candidly reveals the mistakes and lessons learned along the way. His storytelling not only captures the adrenaline of aviation, but also serves as a powerful testament to resilience, learning, and the shared camaraderie of those in the skies.

Join McBride on this remarkable journey and experience the highs and lows of flying like never before.

First published 2014. Second edition December 2024.

FROM THE FLIGHTDECK

More stories from the 'sharp end.'

Following the successful release of The Flightdeck Survival Manual, seasoned pilot and veteran author James McBride invites readers to delve deeper into the realities of aviation through more stories from 'the sharp end.'

Introduced by John Peters, a Gulf War veteran and renowned motivational speaker behind the bestseller "Tornado Down", this collection promises an eye-opening exploration of the skies.

The anthology shares tales of airborne interceptions, the challenges of teaching unteachable pilots, and the dogged pursuit of success in air displays.

Each story immerses you in the highs and lows of a pilot's life, offering candid and inspiring insights. Buckle up and prepare for a ride filled with experiences that you won't find anywhere else.

First published 2016. Second edition April 2025.

FLIGHTDECK PROSE & CONS

Secrets from the little room at the front of the plane.

Candid aviation short stories telling secrets from that little room at the front of the plane - the Cockpit!

The highs and lows of flying airliners all over the world. Technical Faults, Dangerous Weather, Disruptive Passengers and Lucky Escapes.

A "Lucky Engine Failure", another "KISS", an "Emergency Descent, the truth behind the headline making tales is inside.

The backstory to the Demoflight of a refurbished private jet and flying the Prime Minister in a private airliner. Did we miss our ATC 'Slot' out of a busy Heathrow airport with the PM onboard?

Things we NEVER told the passengers. They are all here, told by a highly experienced Senior Airline Captain who has seen it all.

First published March 2021.

HOW TO STEAL A JET

And other aviation stories.

Anybody can steal a jet. Of course, a few tools will make the task easier.

You'll want a good network of contacts, familiarity with the "get-away" airport, a credit card with sufficient balance to fill the tanks, cash to garner help and "goodwill" from airport officials and a lot of brass neck.

Flying the multi-million-dollar loot is the easy bit. Join Captain James in these airborne capers and as he shares behind-the-scenes stories from his 40-year career in aviation. It's all here!

First published June 2024. Copyright © James McBride 2024.

What they said about HOW TO STEAL A JET

"James' best book yet!" – Jeremy Pratt, author and owner of AFE Ltd.

"Unveils the human experience behind the glamour of jet-setting..."

Captain Chris Pohl, Virgin Atlantic
@captainchris

"McBride pulls no punches... A breath of fresh air."

Captain Steve Ford, author of 20 West

ABOUT THE AUTHORS OF AIRCREW STORIES

Independent author and long-time columnist for Flight Training News, **James McBride** draws writing inspiration from a 40-year aviation career. He began with Her Majesty's Royal Navy where he piloted the Harrier and helicopters before taking to commercial flight in the late '80s. Between logging more than 17,000 flight hours in and out of 94 countries on scheduled and corporate jets, McBride trained and tested pilots. He also spent his free time flying vintage aircraft in air displays.

Perhaps best remembered by young pilots he encounters were the years he played himself, "Captain McBride," in the Airline reality TV show. The programme captured behind the scenes activities of UK low-cost carrier easyJet from 1999 to 2007.

"Watching you fly the 737 on the EasyJet show over 20 years ago inspired me to become an airline pilot," wrote one fan. "I'd watch those episodes on repeat....Now I'm a captain at Virgin flying the 787. Dreams really do come true!"

McBride logged his final flight miles before retirement captaining a Boeing 757 for DHL into Miami in November 2021. His writing has been published by Flyer and Flight International magazines and online by Aeropodium Expert. When not at the computer, he enjoys motorcycling, sailing, skiing, diving, and travel. He resides with his wife, Rosemary Macray, in Florida.

Kais Kassim – fell in love with publishing in the 1980s and enjoyed a career spanning over 20 years of enjoyable work on two in-flight magazines for leading airlines where aviation and print came together into his life forever.

Moving on with technology and the Internet, he helped publish several aviation book titles on paper and digital print. He now enjoys working in the corporate jet sector and online training for airline pilots.

Printed in Dunstable, United Kingdom

68149412R00131